Digital Innovations in Architecture, Engineering and Construction

The Architecture, Engineering and Construction (AEC) industry is experiencing an unprecedented transformation from conventional labor-intensive activities to automation using innovative digital technologies and processes. This new paradigm also requires systemic changes focused on social, economic and sustainability aspects. Within the scope of Industry 4.0, digital technologies are a key factor in interconnecting information between the physical built environment and the digital virtual ecosystem. The most advanced virtual ecosystems allow to simulate the built to enable a real-time data-driven decision-making. This Book Series promotes and expedites the dissemination of recent research, advances, and applications in the field of digital innovations in the AEC industry. Topics of interest include but are not limited to:

- Industrialization: digital fabrication, modularization, cobotics, lean.
- Material innovations: bio-inspired, nano and recycled materials.
- Reality capture: computer vision, photogrammetry, laser scanning, drones.
- Extended reality: augmented, virtual and mixed reality.
- Sustainability and circular building economy.
- Interoperability: building/city information modeling.
- Interactive and adaptive architecture.
- Computational design: data-driven, generative and performance-based design.
- Simulation and analysis: digital twins, virtual cities.
- Data analytics: artificial intelligence, machine/deep learning.
- Health and safety: mobile and wearable devices, QR codes, RFID.
- Big data: GIS, IoT, sensors, cloud computing.
- Smart transactions, cybersecurity, gamification, blockchain.
- Quality and project management, business models, legal prospective.
- Risk and disaster management.

Pierpaolo Ruttico
Editor

Coding Architecture

Designing Toolkits, Workflows, Industry

Springer

Editor
Pierpaolo Ruttico
INDEXLAB, DABC
Politecnico di Milano
Milan, Italy

ISSN 2731-7269 ISSN 2731-7277 (electronic)
Digital Innovations in Architecture, Engineering and Construction
ISBN 978-3-031-47912-0 ISBN 978-3-031-47913-7 (eBook)
https://doi.org/10.1007/978-3-031-47913-7

This Springer imprint is published by the registered company Springer Nature Switzerland AG
The registered company address is: Gewerbestrasse 11, 6330 Cham, Switzerland

Preface

In the context of contemporary architecture, it is often said that we are all dabblers. In a sense, this is true, or rather, despite not necessarily being experts in fields like mathematics, computer science, and mechatronics all at once, we are still equipped to tackle the design and construction of highly complex architectural structures. This is made possible by a type of knowledge encapsulated within design and manufacturing tools, and by architects who act as "glue people", capable of integrating different disciplines through coding.

As Andrew Witt writes in his recent book "Formulations": "Today, computational tools that encapsulate mathematical methods are radically short-circuiting the path to expertise, blurring the facile distinction between dabblers and virtuosos, and democratizing access to the systems and aesthetics of mathematical design."

The culture of using computational tools that make the application of mathematics easy is vital for contemporary architecture, but at the same time, it is an extremely fragile culture that can be easily lost. It depends on the commitment of passionate architects and engineers who have transformed from "amateurs" into agents of innovation. They first learned to write software and then integrated it into new hardware, creating both tools and processes that are revitalizing an increasingly "codified" architecture industry.

This book aims to preserve and disseminate some aspects of this delicate "instrumental knowledge".

"Coding Architecture—Designing Toolkits, Workflows, Industry" offers a clear picture of how the Architecture, Engineering, and Construction (AEC) sector is evolving, moving toward an increasingly smooth integration between software and hardware, between algorithmic design tools and numerically controlled machinery.

The book is divided into three parts. The first one features some of the most popular TOOLKITS—Software, plug-ins, add-ons, platforms, frameworks—and their genesis, explained by their creative developers. The second part focuses on virtuous WORKFLOWS and design models, pioneered by top companies worldwide. The third part presents the most innovative construction INDUSTRY, driven by young entrepreneurs and forward-looking companies. Toolkits, workflows, and industry demonstrate their virtuosity when they integrate with each other.

"Coding Architecture" clearly highlights the urgency of integrating design and construction—throughout coding—to make the built environment sustainable and beautiful.

The goal of this book is twofold: to reveal the "state-of-the-art" of the aforementioned topics through cutting-edge case studies, and to present a "vision" of where Architecture is heading, through the personal insights of the authors. It is essential to emphasize that the authors of this book are the real drivers of the radical change that is taking place in contemporary architecture, and they are part of a wide interconnected cultural and technological system.

The authors of the "Effective TOOLKITS" part are the actual developers of some of the most popular software toolkits and frameworks used in the AEC industry today. These are plug-ins or platforms that have radically changed the way the AEC industry operates since 2008 when Grasshopper3D for Rhino (a visual scripting platform for composing algorithms—by David Rutten) broke through to architects. In the past decade, the selected authors, among others, have shaped the history of architecture and engineering. They have hundreds of thousands of followers in the community, usually architects, engineers, and designers, who not only follow them in the evolutions of the field, but also actively use their add-ons and plug-ins. Within the pages of this first part of the book, a dynamic exploration of technological advancement and innovative thought unfolds through five distinct contributions.

In the first chapter, the power of Large-scale Text Generative Models, exemplified by DALL-E and Stable Diffusion, takes center stage, presenting a transformative opportunity to elevate design communication through images. These models, demonstrated through the "Ambrosinus-Toolkit" project by Ambrosini, show the fusion of AI and creative design, offering insights into the ever-evolving realm of artificial intelligence.

The second chapter—by Walters and Bariviera—delves into the vital realm of sustainability within the construction industry. Through Eckersley O'Callaghan's ECO2, a novel approach emerges to assess embodied carbon emissions in building design. By aligning projects with esteemed standards set by RIBA and LETI, the potential to meet and exceed ecological benchmarks becomes evident, promising a sustainable future for the industry.

Software development within the AEC sector takes center stage in the third chapter, thanks to Lombardi from Buro Happold. Open-source solutions such as BHoM and Speckle offer glimpses into the future, where interoperability and collaborative innovation transform architectural practices and redefine industry standards.

The fourth chapter narrates the birth and evolution of HAL Robotics by Schwartz. His journey, from generative design experiments to the development of the HAL Robotics Framework, reshapes the landscape of industrial manufacturing. This chapter stands as a testament to the crossroads of architectural design, software engineering, and automation technologies.

Lastly, urban complexities are explored by Beltracchi, emphasizing resilience and ethical evolution in the face of environmental crises. The narrative navigates the framework of intricate connections between technology, urban structures, and community interactions. The exploration of metaverses and their potential for post-disaster reconstruction intertwines architectural theory, AI, and blockchain, promising a paradigm shift in urban planning.

The authors of the "Virtuous WORKFLOWS" part are pioneers of the contemporary paradigm shift in architecture within leading design firms worldwide. They are architects and engineers who actively use a multitude of software, finding the best way to integrate them with each other. Very often they are the coincident figure of software developers, other times they are users and advise developers to implement tools for the benefit of increasingly integrated and virtuous processes.

Andaloro's "Towards a rhizomatic design model" introduces a paradigm shift by emphasizing interdisciplinary collaboration facilitated by digital tools. This model, as opposed to a linear one, particularly emphasizes sustainability and fabrication.

Baldini, Brisolin, and Cerutti's chapter presents a workflow that optimizes user comfort by balancing daylight, glare, and overheating in buildings. Their iterative parametric approach empowers designers to create optimal facade configurations while considering architectural vision and operational embodied carbon reduction.

The chapter "Testing digital workflows on building renovations" by Sangiorgio, Bogdanova, Globa, Jacob, Vela, and Bianchini explores the transformative potential of digital tools in retrofitting existing buildings. Through a case study of the F10 New Law Building at the University of Sydney, the authors showcase the benefits of using digital workflows for analyzing, reconfiguring, and optimizing structures.

"Computational artwork to transform places" by Fox and Pagani delves into Studio Chris Fox's projects, where computational design processes and digital workflows enable aesthetic experimentation and material optimization. The chapter highlights the coordination challenges involved in delivering complex artworks.

Robeller's chapter introduces the Reciprocal shell, a lightweight timber grid structure designed for efficient fabrication and assembly. Through an innovative "design-to-production" workflow, the project demonstrates the potential of robotic fabrication in achieving fast and precise results. This chapter leads us into the third part of the book.

The section titled "Innovative construction INDUSTRY" brings together contributions from young entrepreneurs and innovative companies that have mastered the challenge of automation for the production of building components and subcomponents. From 3D printing of concrete, clay, metal, and polymers to digital fabrication of highly complex elements, the featured companies operate successfully from small to large scale, demonstrating the possibility of integrating design and manufacturing in a sustainable way.

Scaburri examines the impact of digitization on the construction sector, highlighting the fusion of advanced tools like parametric design and BIM with manufacturing and assembly systems to forge a new digital chain. Large-scale innovative and beautiful projects by Stylcomp are presented.

"Transforming the Construction Industry through Wire Arc Additive Manufacturing" by Costanzi, Waldschmitt, Knaack, and Lange spotlights six research projects, showcasing the versatility of WAAM in AEC, addressing topics such as mass customization, onsite fabrication, and post-processing.

"Custom-made Printheads Empowering 3D Printed Concrete" by Volker and Chris explores the potential of 3D concrete printing developed by Vertico, emphasizing the synergy of robotic technology, material mix, and computational design tools in the larger construction landscape.

Ter Hal discusses how architects could leverage their role to enhance digital workflows and protect creativity in capitalist conditions. Elegant and innovative projects by the Studio RAP are presented by the author.

The last chapter by Ruttico, Bordoni, and Maccagnan introduces the Superframe system, a transformative approach to complex geometries that balances conscious resource utilization and computational design possibilities, achieved through the innovative integration of planar panels and 3D-printed frames.

These chapters, collectively, invite readers on a journey through innovation's frontiers, offering insights into the ever-evolving synergy between design, materials, technology, and human endeavor.

"Coding Architecture" is intended to be a valuable reference book for architects, programmers, engineers, and entrepreneurs, who are passionate about innovation in the AEC industry and wish to have best practice examples and guidance to shape more sustainable and beautiful world.

Milan, Italy Pierpaolo Ruttico

Contents

Part I
Effective Toolkits

Chapter 1
Ambrosinus-Toolkit Plugin: Artificial Intelligence Text-to-Image Generative Models Through Grasshopper

Luciano Ambrosini

Abstract Large-scale Text Generative Models known as DALL-E and Stable Diffusion offer incredible opportunities to improve and enhance the creation and manipulation of design communication through images. The toolkit developed by the author shows how to integrate the LTGMs technology introduced by the OpenAI and StabilityAI platforms into the creative design process as well as the Dense Prediction Transformers (DPT) technology capable to predict a 3D-shaped object started from a 2D image generated by the AI. This contribution will provide the key elements to understand some fundamental aspects of research on artificial intelligence, while through the "Ambrosinus-Toolkit" project, publicly shared on the network, the reader will be provided with operating cues that can be experienced during the schematic design stage. The research presented bears witness to a first step foot with a new paradigm that uses Diffusion Models in the computational design workflow.

Keywords Artificial intelligence · LTGMs · Dense prediction transformers · Diffusion models · Generative AI · Human-AI

1 Introduction

The impressive diffusion of Large-scale Text-to image Generative Models (LTGMs) and the unimagined quality returned by such digital processing exploded between July and August of the year 2022 and polarized the attention of the Information Technology (IT) industry and of the Academies that have invested resources in the R&D on AI-powered products in the last ten years.

The research projects carried out by OpenAI, StabilityAI and MidJourney (MJ) through their proprietary browser-based platforms have received wide acceptance from designers, researchers and AI enthusiasts who have been involved in a large beta-testing process so much to "suddenly" make AI a tool that artists and illustrators

L. Ambrosini (✉)
Indipendent Researcher, Napoli, Italy
e-mail: luciano.ambrosini@outlook.com; info@lucianoambrosini.it

P. Ruttico (ed.), *Coding Architecture*, Digital Innovations in Architecture, Engineering and Construction, https://doi.org/10.1007/978-3-031-47913-7_1

can no longer do without. Consequently, and just as quickly, the AEC industry was also surprised by this important cultural phenomenon, immediately imagining the opportunities for innovation that could arise from it. Even if the concept of "technology transfer" has been a known practice in the architecture, engineering and construction fields since the 80s, since the first AEC-tech conferences of the then-nascent giant Autodesk but not only, in this case, the mainstream phenomenon seems overbearingly manifesting the true symptoms of the fifth industrial revolution—the one based on the cooperation between people and artificial intelligence that will take humans from the digital industry to the digital society. An immediate explanation is provided by the high level of maturity of AI research—at least for specific tasks—and its definitive democratization through the global network.

This contribution aims to bring the reader's attention to the immediate applications in the AEC field resulting from the latest scientific research conducted by the main players in the AI panorama and publicly disclosed through the sharing of Python algorithms and the pre-trained model datasets, all in perfect open-source style through a network of repositories such as Huggingface, GitHub and Google Colab.[1] In order to cast this new technological paradigm within the main design tools, such as the most widespread visual programming language (VPL) platforms like Grasshopper (McNeel and Associates) and Dynamo (Autodesk Inc.), a specific toolkit has been developed in the Grasshopper (GH) environment so as to overcome the current limitation which gets these functions available only for applications browser-based or as a command-line interface. Through the development of specific nodes, it is possible to perform some of the most interesting functions based on AI text-to-image and image-to-image generation. Moreover, the possibility of testing with this toolkit a simplified experimental procedure capable of exploiting the monocular depth estimation throughout large-scale pre-trained datasets to reconstruct 3D objects as mesh geometry starting from a single-shot image (2D image).

The OpenAI and StabilityAI platforms were deemed particularly suitable for the stated purpose due to a greater propensity of these companies to publicly release the Application Programming Interfaces (APIs)—a set of definitions and protocols for creating and integrating applications software that allows interaction with DALL-E and Stable Diffusion. In any case, the strong acceleration impressed by MJ is recognized with no doubt, above all from the second test phase which began in April 2022 and in particular during the transition between v3 and v4, July and November 2022, where the spread of high-impact artistic image production based on text-to prompt generation has reached a large audience.[2]

[1] Google Colab or "Colaboratory" is a browser-based programming platform that allows students, researchers and simple users to execute Python code and exploit GPU hardware remotely for free or by subscription. Google Colab Homepage, https://colab.research.google.com/, last accessed 2023/01/31.

[2] Midjourney works basically as a Discord bot, an automated module within the instant messaging platform designed for communication between gamer communities and internet users. MJ is a tool best known for its artistic-style results.

In terms of "medium", the choice was Grasshopper as a capstone of the AEC industry regarding computational and parametric design, a platform perfectly integrated into the architectural and engineering generative processes. Its node logic[3] has cleared in a particularly effective way two of the fundamental aspects of nowadays digital design: the integration of programming languages in the architectural design practice and the "community" aspect through the practice of crowdsourcing applied to coding and to scripting. Through the Grasshopper's community, it is as if the conditions of Marshal McLuhan's global village were re-proposed in a modern key in which all the contrasts of ideas and the varied possibilities of accessing widespread knowledge were fully exploited. Through this two-way experience, the comparison of knowledge becomes a productive action for everyone and feeds a vast market of plugins which, programmed by individual users or research teams, greatly amplify the functional capabilities of the platform, extending its use to multiple disciplinary fields. "Ambrosinus-Toolkit", is the name of the developed resource, which aims to extend some of the main features offered by the OpenAI and StabilityAI platforms to the vast world of computational designers who adopt Grasshopper every day in their creative and design processes.

2 From AI Key Concepts to the Large-Scale Text-to-Image Generative Models (LTGMs)

A branch of artificial intelligence research well known in the AEC industry is undoubtedly that of Machine Learning (ML). This branch uses statistical techniques to give the machine the ability to "learn" from the data without explicitly receiving instructions on how to do it. This process is known as "training" a model using a "learning" algorithm that progressively improves the model's performance on a specific task. In the scientific literature, there are four main categories of approach to ML also called "Five Tribes" [1], in particular: *evolutionaries*, *Bayesians*, *symbolists*, *analogizers*, and *connectionists*—Alan Touring, for instance, can be counted among the symbolists with his test discussed on "Computing Machinery & Intelligence" in 1950 making him the fundamental forerunner in research on artificial intelligence supported by computers—although the term AI was officially coined by John McCarthy in 1956. Nevertheless, ML research is not so rigid in defining these approaches in terms of theories and principles of inference, which is why researchers over time have used mixed and hybrid approaches which have defined the so-called SOTA (State of The Art), best practice for specific tasks.

Architectural and engineering design firms experiment with ML every time they resort to the use of computational tools, especially in the optimization processes

[3] The "node" logic is a user interface simplification gimmick already known in the Maya Hypergraph language. In the early 2000s, the lack of agreement with Bentley Systems gave way to the development of Grasshopper (David Rutten, 2008) strongly detaching from the GenerativeComponents project (John Nastasi, 2003) forerunner of VPL platforms.

of the elements that articulate the architectural project with purposes ranging from the structural to the environmental fields, helping them to define shapes, behaviours in short making designed decisions. For instance, tools such as Galapagos, Goat, Opossum, SilverEye, Octopus, Wallace etc. allow the use of advanced functions of the single-objectives or multi-objectives type, through the adoption of algorithms based on genetic evolution (GA), Particle Swarm Optimization (PSO), Ant Colony Optimization (ACO) etc., others instead offer the possibility to choose which type of algorithm to perform optimization with, this is the case of Opossum among those mentioned. The latter one includes a model-based RBFOpt and evolutionary CMA-ES. It also includes the multi-objective RBFMOpt, and the multi-objective MACO (Ant Colony), MOEA/D, NSGA-II and NSPSO (Particle Swarm).

About Grasshopper over time several projects, more or less well-known, have been implemented to slowly introduce ML functionalities that expand the platform's capabilities. Among the best-known projects is LaunchBoxML by Provingground (author Nathan Miller)—a generalized Machine Learning toolkit for implementing models including regression, Neural Networks (NN), probabilistic classifiers, and clustering.

An example of non-linear regression was designed by the author for i-Mesh Company in regards the membrane shading system of Expo Dubai 2021 [2]. Among the projects still under development (such as Owl by Mateusz Zwierzycki, DoDo by Lorenzo Greco, Crow by Benjamin Felbrich, Tunny, Amoeba BESO and tOpos), the possibility of introducing the Google TensorFlow (TF) library to implement various computational tasks is particularly interesting like the well-known example of the plugin named "Pug" by Diego Apellániz. This library was developed by the Google Brain team initially for internal use only and subsequently made public. TF supports a wide range of tasks, including ML, Deep Learning, Image Recognition, Natural Language Processing, and Speech Recognition. Deep Learning (DL) is an important mechanism for understanding the evolution of AI and above all allows the introduction of a further key concept that takes us into this historical period that has cleared the LTGMs, namely the ability to process different nature billions of data. DL is a subset of ML that attempts to mimic the neural learning activity of the brain to recognize complex patterns in the data. The word "deep" refers to a large number of information "levels" that help achieve better performance by mining knowledge from huge databases. Definitively through the DL is possible to increase the solving capabilities of the so-called Artificial Neural Networks (ANN or NN) a computational model as an engineering simplification of the human neural system. The word "suddenly" in the use of the AI expressed in the introduction was deliberately used to underline the strength and the maturity of the noticeable results that have been condensed, as known, in decades of research starting from 1936 precisely with Alan Turing. Their success is certainly not attributable to the recent media clamour much less to the commercial one. The initial "hardware" difficulties encountered by the researchers have disappeared over time and the results were not long in coming. The last fifteen years have been decisive in improving the DL and pushing forward the technological progress in hardware manufacturing thanks to the enhancement of the Graphic Processing Units (GPUs) specially led by NVIDIA company. As early as 1997, technological enhancement was decisive in the victory of IBM's Deep Blue system over the chess champion Gary Kasparov. He underlined that the real added value would be to pursue a "man plus machine" approach that would guarantee more sustainable results over time. The successes of AI have jumped to the news thanks to the scientific results achieved by DeepMind, an artificial intelligence company founded in September 2010 in London, UK. It was acquired by Alphabet Inc (Google's parent company) in 2015. DeepMind's main focus is on developing advanced AI algorithms and systems to solve complex problems in fields such as gaming, healthcare, and climate modelling.

A significant example in terms of AI applied to the architecture field is the introduction of a particular type of DL, the Generative Adversarial Network[4] (GAN), invented by Ian Goodfellow in 2014. The GAN has literally paved the way for generative models (creation and interpretation of images), which in architecture has translated into AI-based generative design. Researcher Stanislas Chaillou, University of Harvard, points out that the determining factor was to transform the GAN from an analytical tool to a generative agent [3] and this turned it into an interesting design tool. Different releases of GANs have been developed, such as StyleGANs and CycleGANs. Basically, each version has been a valuable improvement to achieve the ability to generate images of non-existent buildings which reflect a series of stylistic criteria, a kind of morphogenesis raster-type-based coming from the input data only.

One of these versions, called ArchiGAN, was implemented and disclosed by Chaillou [4]. ArchiGAN is a generative adversarial network designed for architecture. It is a type of AI model that is trained to generate architectural designs or architectural elements. Given a set of input parameters, such as building style, building height, and floor plan size, ArchiGAN can generate a corresponding architectural design. This technology has the potential to revolutionize the architecture and construction industry by allowing architects to quickly generate and visualize new designs, reducing the time and cost associated with traditional design processes. Additionally, ArchiGAN can also be used for educational and research purposes, providing students and researchers with a tool to experiment and understand architectural design principles and theories. Another key concept that brings us closer to LTGMs technology is undoubtedly the contribution of research on "language", specifically Natural Language Processing (NLP). The first studies date back to the 1950s, such as "The Logical Structure of Linguistic Theory" by Noam Chomsky, published in 1955, then followed by the studies of John McCarthy's research group until the 1970s with "A Program that Simulates Human Thinking" by Roger Schank and Robert Abelson, published in 1977, which introduced the concept of script-based reasoning and demonstrated the feasibility of NLP models for natural language understanding. Through the DL, research in the field of NLP undergoes a significant boost thanks to a multimodal approach in processing data of different natures (not only text but also voice and image). At this point the assumption of the researchers Jaime

[4] GAN are two neural networks: a generator that makes new data and a discriminator that tells real data from fake data. They are trained in a game where the generator tries to fool the discriminator and the discriminator tries to catch the generator.

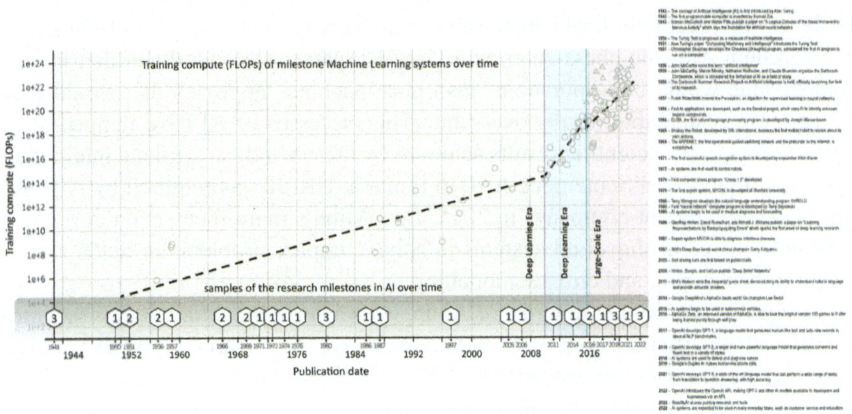

Fig. 1 Three eras of the Machine Learning and some main AI research milestones (*source* author)—a high-res image is available at this link [6]; data source from Sevilla and Heim [5]

Sevilla and Lennart Heim is clarifying through which it is possible to define three Eras of the ML using the advent of the DL (2010–2012) as a discriminating factor [5]: Pre-DL Era, DL Era and Large-scale Era (see Fig. 1).

The 2022 state of the research on AI documented by Stateof.ai [7], discloses the following data: in the Pre-DL Era the "training" speed doubled every 20 months, in the DL Era (2010–2015) it doubled every 6 months and in the Large-scale Era (2016–present), data processing quantity capacity first jumped from 100 to 1000 times and then doubled every 10 months in terms of speed. It could be said that the technological process is definitively abandoning the trend of Moore's Law to make Kurzweil's Law, the "Law of Accelerating Returns" more and more plausible in terms of tech power and social change [8]. Without being strenuously technical, practically this law provides for the achievement in the tech-progress timeline a point of "singularity", a condition where AI will surpass Human intelligence. Probably one of the valuable social goals of AI research is teaching humanity how better to know itself and perhaps this will happen when, and if, Kurzweil's prediction comes true. The AI progress downside is the computationally intensive and time-consuming cost of training a large-scale model from scratch, which is why the so-called "pre-trained model" is often used. Initially, an AI model is first "trained" for a generic task using an initial training dataset through a process called "pretraining" (first step). The resulting model is subsequently retrained using another specific training dataset for the second pre-selected task, and usually, this set is much smaller than the pretraining. These specific activities are called "downstream" (second step) and the retraining process is called "fine-tuning" (third step). The knowledge of the original model is then used for a more specific task. Therefore, this process is called "transfer learning". Ultimately, the basic idea is mostly hedonistic, that is, to reuse the knowledge already learned to solve new problems. Therefore, the original model is referred to as the "foundation" or "pre-trained model". Pre-trained models can help accelerate AI application development, as they allow developers to leverage the knowledge and work of others to gain an edge on their own projects. They are commonly used in computer vision and natural language processing, and there are many models publicly available for a variety of tasks, such as image classification, object detection, and sentiment analysis. Pre-trained model and downstreaming are, here, the last two key concepts in AI research that open up the use of LTGMs in real-time through browser-based platforms or applications that act as intermediaries in the "prediction" process, which stands for finding patterns and relationships in input data and then using them to predict future events, make decisions, automate processes and provide valuable insights.

2.1 OpenAI GPT-3 (ChatGPT)

OpenAI is a leading artificial intelligence research organization that was founded in 2015 by Sam Altman, Reid Hoffman, Jessica Livingston, Elon Musk, Ilya Sutskever, Peter Thiel and others. It was a non-profit until 2019 then has become OpenAI LP. The goal of OpenAI was to promote and develop friendly AI that would benefit humanity as a whole, and to ensure that AI technology was developed in a responsible and safe manner. From their first announcement, they said: "We believe AI should be an extension of individual human wills and, in the spirit of liberty, as broadly and evenly distributed as possible. The outcome of this venture is uncertain and the work is difficult, but we believe the goal and the structure are right [9] […]". Indeed, it has been involved in a wider range of AI-related topics and initiatives, such as safety and security, transparency and accountability, and ethical considerations.

Since the early years, OpenAI focused on developing cutting-edge research in AI, including areas such as Deep Learning (DL), Reinforcement Learning (RL), and generative models (GMs). RL is another key concept of AI research and is one of the three learning modalities[5] adopted. In particular, RL is an area of ML in which software agents learn goal-oriented behaviour by trial and error in an environment that provides rewards or penalties in response to their actions.

In this regard, one of the research initiatives made available by OpenAI is a sort of "Roman gymnasium" named Gym, i.e., a gym which allows developers to create virtual environments to train AI agents; Gym carries two other initiatives: Universe, which allows developers to use games and other applications as training environments; and Baselines, which provides a suite of reinforcement learning algorithms. Gym is a tool accessible through official channels and allows to train of new models through the use of specific publicly shared Python libraries. However, this company has become particularly famous as it is actively involved in the development of Generative Pretrained Transformers (GPT) and the widespread DALL-E model. GPT is a

[5] The other two modes are the Supervised Learning (SL), that uses labeled data to train a model and the Unsupervised Learning (UL), that uses unlabeled data to find patterns and structure. Anyway, exist also another mode defined as Self-supervised learning (SSL), that is a kind of UL that creates artificial labels from raw data.

transformer-based[6] language model that uses deep learning to generate human-like text. It was pre-trained on a large corpus of text data and can then be fine-tuned for specific NLP tasks such as question-answering, text completion, and summarization. Even if the first developments date back to the DL Era (around 2010), version 3, the most famous and currently the subject of numerous research papers, is the GPT-3 made public in June 2020. This version has increased the number of input parameters that can be processed going from 1.5 billion of the GPT-2 version to about 175 billion of the current version [10]. The consumer world is learning to use this powerful tool through the OpenAI platform called "ChatGPT". Thanks to its capability of generating human-like text, this technology will enable a huge implementation of multiple tools for production and intellectual work. A particularly coveted feature is to generate or complete pieces of code written in different programming languages. In this sense, in January 2022 a variant of GPT-3 called InstructGPT (GPT-3.5) was developed which is able to improve the reliability and use of the text-to-code generation function. However, in February 2023 the company provided a tool [11] called "AI text-classifier" to verify the authenticity of texts actually written by a human and not by an artificial intelligence. This mitigating measure was necessary due to improper use of this tool deemed harmful to the students learning process of New York City public school [12]—however, it is not too difficult to extend this effect to various educational realities, not only Americans, so as to point to it as a misinformation bot. Even Wired magazine, which has always dealt with technological progress, has considered this tool a historical event worthy of note, so much so as to compare it to the introduction of the World Wide Web into the global cultural panorama. However, Sam Altman, who co-founded OpenAI with Elon Musk, later thought about diluting the so-called hyping, as reported by the MIT Technology Review [13]: "The GPT-3 hype is way too much. It's impressive [...] but it still has serious weaknesses and sometimes makes very silly mistakes. AI is going to change the world, but GPT-3 is just a very early glimpse. We have a lot still to figure out". In any case, the field is totally expanding in the coming months, for instance Google will introduce its GPT chat, LaMDA a Language Model for Dialogue Applications [14] with the name of "Bard" and its probable integration into all its productive tools as well as in its renowned search engine. The experimentation and the usage of this tool at the moment seem to bend only to the limits of the creativity of professionals of any scientific-technological sector projected towards the future.

2.2 Diffusion Models: DALL-E and Stable Diffusion

Large-scale models empowered by the use of linguistic models such as GPT become a mainstream phenomenon as anticipated in the introduction and do so through

[6] Transformers, introduced in 2017, are a type of neural network for processing sequential data such as text. They use a self-attention mechanism to understand the meaning of words in context by weighing their importance in the input sequence.

two Neural Networks (NN) with a prevalent Bayesian approach, called DALL-E—crasis between the name of the famous artist Salvador Dalì and Wall-E, the animated character of Pixar—and Stable Diffusion, this latest open-source from the StabilityAI organization. They are both tools capable of generating incredibly sharp images by inputting descriptive text and images. Both belongs to the macro category of Diffusion Models, AI models which through the literal description and images provided as input, carry out a denoising process [15] by means of a NN, defined Decoder, trying to reconstruct an image that is more and more coherent with the given input. The first phase of this process requires the pre-training of the neural model. The first real phase is realized by a different NN, defined Encoder, able to reduce/transform each labelled-text and image input into vector information, thus fragmenting each image into numbers. Basically, the image gets dirty through a noise effect called the *corruption process* that the Decoder will "clean up" afterwards. The input data are defined within a mathematical space called "latent space" within which the Decoder will "fish" the most reliable solution—extremely simplifying it is as if we wanted to find in a Cartesian space the position of a point through its coordinates (x, y), this is just to give a faint idea because the substantial difference is these models are dealing with billions of parameters.

DALL-E 2 (April 2022) generates images in a two-stage process, first by generating the "essence" of an image and then by filling in the remaining details to obtain a realistic image. The capture of the essence is the result produced by the Encoder, defined CLIP, through contrastive training (labelled-text and images) whose generated image is called CLIP-image. The second step is represented by the Diffuse Model named GLIDE [16], the Decoder that tries to obtain an image defined by the CLIP-image and for this reason, the result produced from time to time is generically named UnCLIP-image [17]. The process is therefore iterated for a certain number of pre-established steps during which the CLIP Encoder tries to "prioritize" the coherence between the produced image and the input text (Fig. 2).

SatbilityAI launches Stable Diffusion (SD) in August 2022, its Diffusion Model is similar to DALL-E and Google's Imagen di Google. As for the previous neural model, SD can perform two particular text-to-image techniques called outpainting and inpainting [18]. The first one is the technique to fill out or extend the area around an image, the second one fills in the missing areas of an image. As mentioned, these operations of Diffuse Models have costs and limits related to the computational

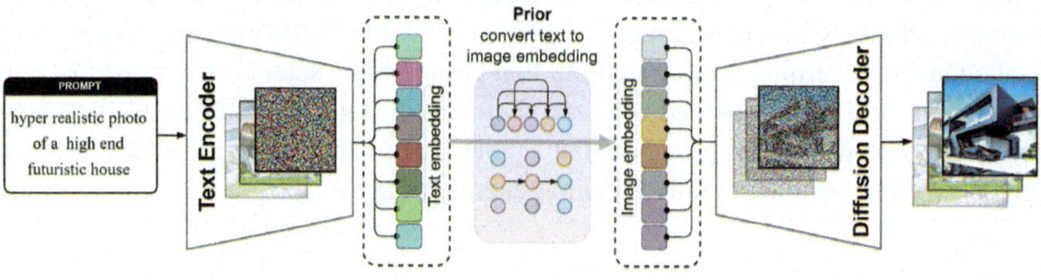

Fig. 2 DALL-E text-to-image simplified process schema (*source* author)

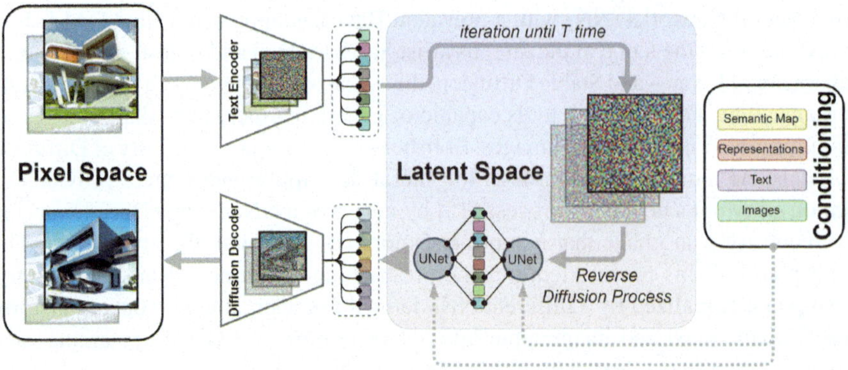

Fig. 3 Stable diffusion text-to-image simplified process schema (*source* author)

capacity of the hardware, for this reason, SD developed the "Latent Diffusion Model" (LDM) thanks to the research group of the Ludwig Maximilian University of Munich [19] which shifts the diffusion process to the latent space level instead of the pixel level. This is based on the notion that the latent space holds most of the semantic and essential information about an image and its related caption, whereas the pixel level contains extra perceptual details which are not necessary. Below, in Fig. 3 a simplified schema of the LDM.

The UNet neural model is a Convolutional Neural Network (CNN) which carries out what is technically defined as segmentation of the images—the recognition of the contents of the images by subsequently attributing to them a "weight" in relation to the consistency between text-image and input prompt through a conditioning process. Both DALL-E and SD have trained their respective encoding processes through a huge number of datasets of images-captions, however, OpenAI has not shared its own dataset consisting of over 400 million images to train its CLIP model classified, which means heavily curated to avoid ethical issues; SD has fully made public its training dataset. The SD has been trained on the LAION-5B dataset which has over 5.85 billion images. It is a Large-scale Artificial Intelligence Open Network (LAION), provides by a German non-profit organization with a stated goal "to make large-scale machine learning models, datasets and related code available to the general public from the non-profit". SD training dataset even if it is larger than the OpenAI one, thanks to the LDM approach can generate equally high-definition images from text prompts relatively in a short time. The launch of Stable Diffusion caused quite a stir in the machine learning world. Up until that point, only a select few companies with vast computing resources had released their text-to-image models, and they were only available with a fee attached, but Stable Diffusion changed all that, allowing anyone with a consumer-grade GPU to generate images.

3 The Development of a Toolkit to Perform LTGMs and DPT-to-3D Technologies Inside Grasshopper

The development of the "Ambrosinus-Toolkit" (ATk) stems from a need for research and application aimed at extending the computational approach to design and to architectural projects by sharing tools with the community of designers and researchers. As announced in the introduction, the computational design platform that has been established as a must-have in professional practice is Grasshopper (GH) and for this reason, it was decided to expand its functionality by exploring two types of AI-powered tools especially: LTGMs and Dense Prediction Transformers (DPT), the latter is a type of AI model that use transformer architecture to process images.

The development of ATk coincided with the greater diffusion of the OpenAI and StabilityAI neural models, more precisely between October and November 2022 following the public sharing of the respective APIs—For this reason, it is necessary to enter the respective API keys which can be obtained through appropriate registration to aforementioned services currently (2023/02/07) in the beta phase. These keys can be used within the developed Toolkit.

Notoriously GH plugins come in the form of toolbars and are generally developed in the programming languages C-Sharp (C#), Python (Py) and sometimes in VisualBasic (VB), however, the procedure adopted for ATk is considered hybrid. The main IT development takes place through the coding of the algorithms in C# using the Microsoft VisualStudio platform an IDE that allows the use of the .NET Framework, and the Rhino 7 and Grasshopper development module provided by McNeel and Associates. The hybrid term refers to the possibility of integrating into ATk both components (also called flow-chart nodes) traditionally assembled in C# and distributed in the proprietary ".gha" format and cluster-type components in the ".ghuser" format. Resorting to codified clusters has had several advantages during the development period which took place parallel to that of diffusion of DALL-E (v.2) and SD. Firstly, the possibility of resorting to the use of the Py language, therefore to the countless libraries developed specifically for the use of AI tools but not only; secondly the possibility of programming the cluster in a modular way and adapting it each time to the feedback received from users and, finally, of making their management and distribution more agile through the GitHub repository platform, guaranteeing maximum diffusion on the network through the author' official channels. Bringing the power of artificial intelligence to the creative design process of architects and designers, which will be shown in the next paragraphs, has raised ATk as the first experimentation has integrated and leveraged the OpenAI, StabilityAI and DPT features inside the McNeel VPL platform. ATk in the current development phase has 5 subcategories for its components, they are About, Display, Image, AI, and Extra. This contribution will mainly focus on the Image and AI subcategories, it should also be noted that the tool present in "About" allows the user and designer to keep track of the entire development timeline of the toolkit through its "changelog", a report about all notable changes made to the project and to its components (Fig. 4).

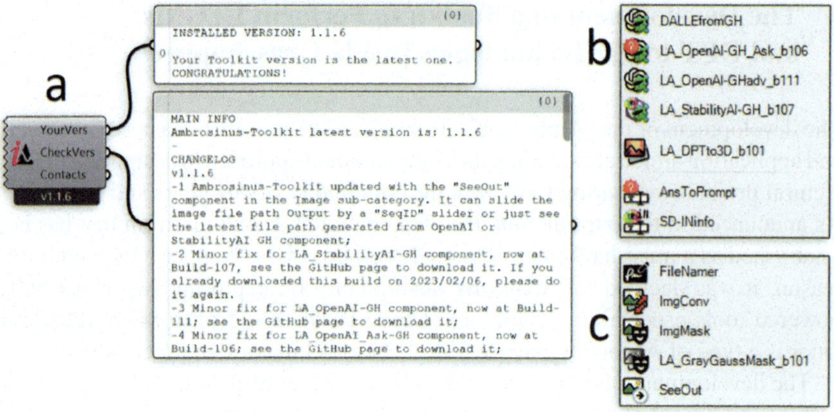

Fig. 4 **a** Info component output; **b** current list of the AI components; **c** current list of the Image components

All the AI components execute a sort of query action, which consists in asking for an AI prediction querying the DALL-E and SD servers. Intuitively, and with the information previously provided, the reader will understand that the prediction process can be divided into two parts: training and inference. During training, the model is fed labelled training data and the model's parameters are updated to minimize the prediction error. During inference, the trained model is used to make predictions on new, unseen data—AI ATk tools act exactly at this stage.

3.1 AI "Prompt-to-Image" Core Components

Most of the AI components developed have been codified in Python language, with the exception of "DALLEfromGH" which is developed in C# and allows the inference action through a special library, "Newtonsoft.Json.dll" because of using the "json" code format. Regarding the other components, the "CPython" tool by Mahmoud M. Abdelrahman was used to overcome the limitation in importing Python libraries defined as non-standard, which actually depends on the IronPython version supplied by default in GH. This limitation will be overcome with Rhino v8 thanks to the Grasshopper component named "RhinoCode (RC)". The last requirement to run these tools is the installation of some specific Python libraries on the Windows OS side, in particular: "openai" and "stability-sdk". This phase has been simplified recently, providing a single file,[7] "requirements.txt", through which in a single event

[7] Required Python libraries: atoolkitdpt, opencv-python, Pillow, numpy, matplotlib, transformers, torch, torchvision, imutils, open3d, openai, stability-sdk, timm. The documentation is reported at, https://bit.ly/LA-WYSIWYTfromDPTto3D and will be publicly shared through ATk v1.1.8/v1.1.9.

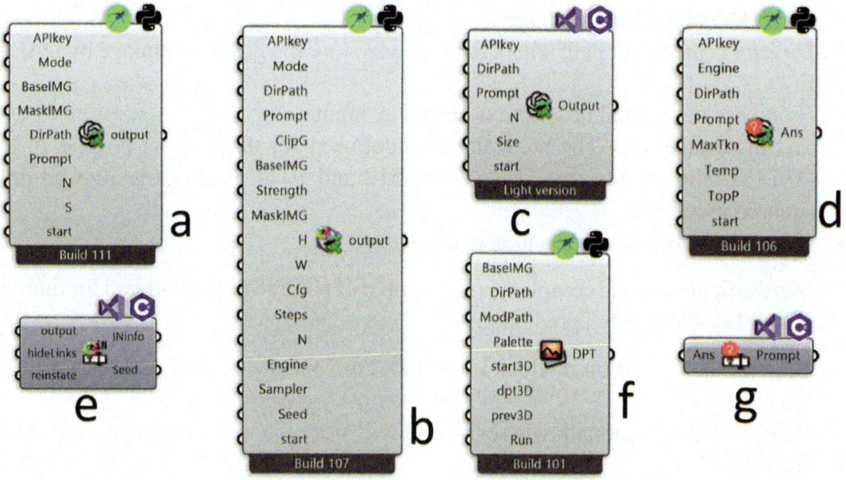

Fig. 5 **a** "OpenAI-GHadv" component runs DALL-E model; **b** "StabilityAI-GHadv" component runs stable diffusion model; **c** "DALLEfromGH" components; **d** "OpenAI-GH_Ask" component runs GPT-3 model; **e** "INinfo" component for SD output images; **f** "DPTto3D" component for depth image estimation and 3D output; **g** "AnsToPrompt" component purges answer from question

it is possible to acquire all the necessary libraries in order to run correctly all the ATk features. Below in Fig. 5 the AI components:

The advanced tools (a) and (b) in Fig. 5 (excluding (c) as a light version of DALL-E) are the AI components cores and are able to perform three different operating modes providing text and images as input, namely: Creative mode, Variation mode and Edit mode. The first mode allows the generation of one or multiple images by passing a descriptive text as input, the second mode returns a variation of the content of the source image, and finally, the third mode allows users to modify part of the source image through an image-mask. The first substantial difference between the two neural models is that in the current version of DALL-E (v2 2022/2023), it is not possible to discretize the generative process through the input parameters, therefore the generated images are subject to a certain randomness; at the same time, the Stable Diffusion neural model is instead able to uniquely identify the parameters supplied as input and associate them with a specific *Seed* value. This uniqueness makes it a very powerful tool and particularly useful if integrated into the creative and design exploration processes because keeping track of the outputs places the designer at the centre of the generative process. Below is a brief list of the main input parameters implemented in the two tools including those necessary to perform the inference process through the APIs of the two selected neural models.

"OpenAI-GHadv" main input parameters:

- *Mode:* execution mode selector (Create, Variation and Edit);
- *BaseIMG* is the source image path, it is required in order to run the Variation and Edit modes;

- *MaskIMG* is the source image-mask path;
- *DirPath* is the target path where the tool will store the generated image by DALL-E;
- *Prompt* is the textual description passed as input;
- *S* is the image size. The pre-trained model used for the training process of the DALL-E allows only these sizes: 256, 512 and 1024 pixels (it is allowed only squared pictures);
- *N* is the number of images to generate;

More articulated, and complex, is the number of parameters considered for running the "StabilityAI-GHadv" prediction:

- *Mode* is the execution mode selector (TXTtoIMG, IMGtoIMG and IMGtoIMG Masking);
- *DirPath* is the target path where the tool will store the generated image by Stable Diffusion;
- *Prompt* is the textual description passed as input;
- *ClipG* is the CLIP guidance mode. It is a tricky procedure executed by the neural network encoder to increase the consistency of the image with the text given as input;
- *BaseIMG* is the source image path, it is required in order to run the IMGtoIMG mode;
- *Strength* is how much "weight" has the text prompt in relation to the initial image (admitted values from 0.0 to 1);
- *MaskIMG* is the source image-mask path;
- *H* and *W* are the height and width of the output image. Only the size in the range 256–1024 pixels with 64 px as increment value is admitted;
- *Cfg* scale dictates how closely the engine attempts to match a generation to the provided prompt; v2-x models respond well to lower CFG (e.g.: 4–8), whereas v1-x models respond well to a higher range (e.g.: 7–14);
- *Steps* affect the number of diffusion steps performed on the requested generation.
- *N* is the number of images to generate;
- *Engine* current version includes 8 engines,[8] all selectable by the user;
- *Sampler* is the sampling engine to use. Currently have been implemented in the toolkit 9 samplers.[9] They are a sort of statistical samplers that are used in the diffusion model prediction process (especially for the denoising process);
- *Seed* is an integer number useful to discretize all parameters used in the generative process. The toolkit assigns a random value if no slider is connected to the *Seed* input.

The second substantial difference between the two neural models is in performing the following equivalent processes: the Variation and Edit mode in DALL-E and the

[8] The Stable Diffusion engines included in ATk are: v1, v1.5, v2, v2.1, inpainting v1 and inpainting v2.

[9] Sampler models included in ATk are: ddim, ddpm, k-dpm 2, k-dpm 2 ancestral, k-dpmpp 2m, k-dpmpp s2 ancestral, k-euler, k-euler ancestral, k-heun, k-lms.

IMGtoIMG and IMGtoIMG Masking in Stable Diffusion. Using "OpenAI-GHadv" it is possible to leave DALL-E to make changes to the source image, but it is not possible to take into account any text prompts; in Edit mode DALL-E will simultaneously consider a source image, an image-mask and the text prompt. The use of "StabilityAI-Ghadv" allows the matching of a text prompt and a source image in IMGtoIMG mode but also guides the output image through the *Strength* and *ClipG* parameters (balancing the weight between the source image and text prompt); in IMGtoIMG mode Masking Stable Diffusion considers a source image, an image-mask, the text-prompt, *Strength* and *ClipG* parameters—fundamental at this stage is to set the same *Seed* of the source image in order to make variations through the mask as described in the text. In this phase, component (e) in Fig. 5 is very helpful for extracting useful image metadata.

Finally, the component (d) in Fig. 5 allows interaction with the OpenAI GPT-3 neural linguistic model. The tool activates the non-continuous Questioning&Answering (Q&A) mode. This mode lets the designer interact through natural descriptive language and obtain answers formulated with equal naturalness (human-like answers).

"OpenAI-GH_Ask" main input parameters:

- *Engine*: current version includes 13 engines, da vinci (3 text versions), ada, curie and babbage, all selectable by the user;
- *MaxTkn*: is the maximum number of tokens to generate in the completion process. It represents the token count of the prompt plus max tokens allowed by the selected engine;
- *Temp*: sampling temperature to use (admitted values from 0.0 to 1). Higher values mean the model will take more risks, which stands for more creativity;
- *TopP*: is an alternative method for sampling, it is similar to temperature (admitted values from 0.0 to 1), it is also called nucleus sampling and represents the amount of the probability mass considered in generating consistent answers.

The reader will understand that the extensive technical statement given above is necessary in order to simultaneously highlight the strengths and weaknesses, the limitations, of these tools simply by evaluating the currently input parameters permitted. Finally, the possibility of documenting the way these tools are used in daily activities is taken into account. The (a), (b) and (d) components in Fig. 5 generate text files and a logfile in ".csv" format in order to let the designer analyse the evolution of the predictions through data storytelling,[10] a task often delegated to the computational designer (Fig. 6).

[10] Video demo, https://youtu.be/1VXe-R44nuw, last accessed 2023/02/13.

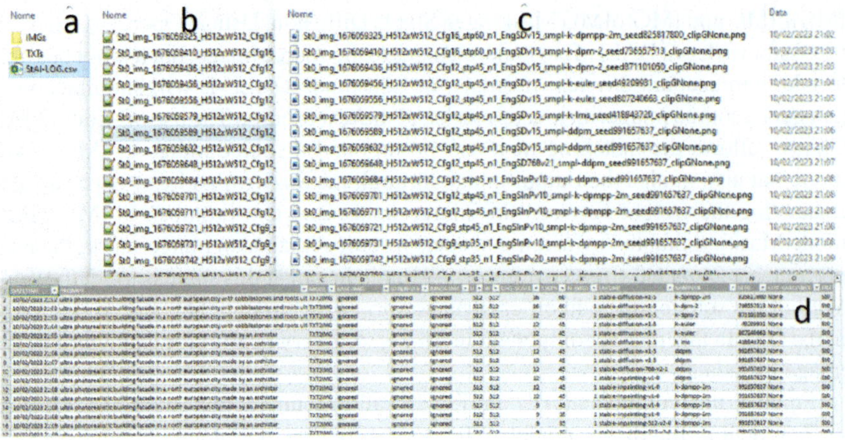

Fig. 6 **a** Target directory where metadata have been stored; **b** text files inside TXTs folder; **c** image files inside IMGs folder; **d** metadata generated by AI tools stored in a unique CSV file to track their usage

3.2 Image Core Components

Both neural models taken into consideration require the input of an image-mask in the Edit mode. Generally, this mask can be an image in grayscale, black/white and a ".png" image in RGB scale and with a transparent cut-out in the case of DALL-E.

In Fig. 7 are shown all the main components.

The adopted workflow requires the use of the ImgMask component for DALL-E and the use of the GrayGaussMask component in the case of Stable Diffusion.

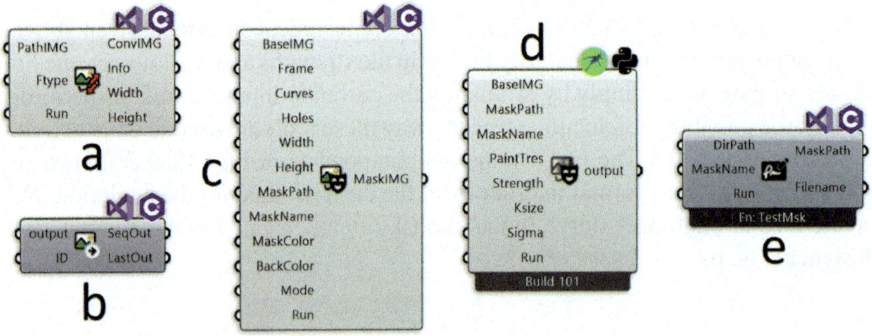

Fig. 7 **a** "ImgConv" component can convert the image in a different format and read generic info about the image's metadata; **b** "SeeOut" component manages the output images; **c** "ImgMask" generate image mask; **d** "GrayGaussMask" component generates image mask; **e** "FileNamer" component helpful in managing the image's filename

The main features of component (c) in Fig. 7 concern the ability to draw polylines within the NURBS Rhinoceros environment by McNeel and Associates to literally shape the image-mask to be subsequently passed as input to "OpenAI-GHadv". The parameters *Mode*, *MaskColor* and *BackColor* help the designer with defining the image output as described at the beginning of this paragraph.

On the other hand, Component (d) in Fig. 7, returns as output a copy of the source image but altered in its colour range, i.e., in grayscale or black/white. This component through the use of a specific Python library called *torchvision* is able to apply a Gaussian blurred effect to the source image in order to make the application of the mask smoother during the execution of the IMGtoIMG Masking mode of the component "StabilityAI-GHadv". The *PaintTres*, *Strength*, *Ksize* and *Sigma* parameters allow the designer to manage the final effect that the output image-mask will have. Once again, it should be underlined, Stable Diffusion has greater flexibility in assigning an image-mask as input and both (c) and (d) components work fine with it.

Finally, components (a), (b), and (e) in Fig. 7 mainly has the scope of simplifying the management of the workflow such as facilitating the extraction of information from the generated images, generating the name of the image files avoiding overwrites and to make easy the sliding preview of the images displayed on the Grasshopper canvas.

3.3 Dense Prediction Transformers (DPT): Depth Map Estimation Inside Grasshopper

Dense Prediction Transformers (DPT) are a type of deep learning architecture used in computer vision tasks such as semantic segmentation, object detection, and instance segmentation. The main idea behind DPT is to predict dense labels for an image by encoding both global and local context information. DPT typically consists of two main components: a *backbone network*, which extracts features from an input image, and a *prediction head*, which makes predictions based on the extracted features. The backbone network is often a pre-trained Convolutional Neural Network (CNN) that has been trained on a large dataset for image classification. The prediction head is then trained to perform dense prediction tasks such as semantic segmentation. Semantic segmentation is a computer vision task that involves dividing an image into multiple segments, or regions, and classifying each region into a semantic category, such as sky, road, person, building, etc.; for instance, it is used in various real-world applications, such as autonomous driving, medical imaging, and satellite image analysis. This neural model was introduced in March 2021 and in particular by Intel researchers René Ranftl, Alexey Bochkovskiy and Vladlen Koltun who managed to recalibrate, with a certain increase in reliability, the capabilities of the DPT on a series of pre-trained datasets with small dimensions in order to carry out two types of analysis, *the monocular depth map estimation* and the *semantic segmentation.* Leaving aside

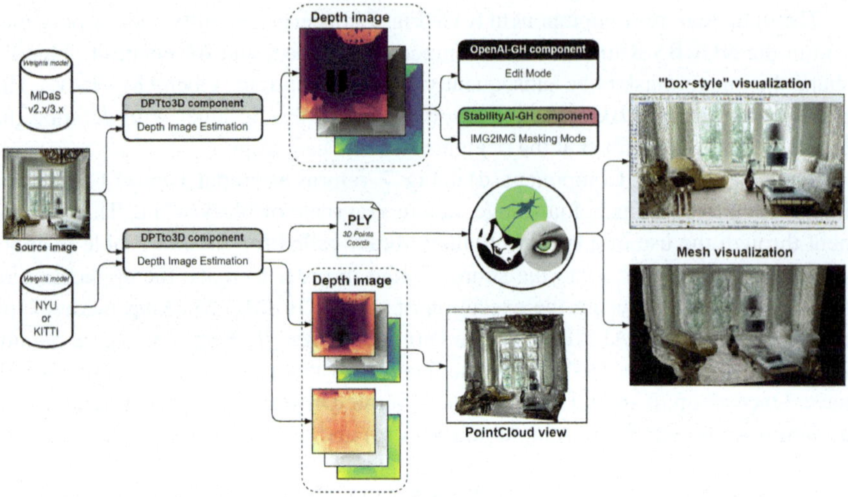

Fig. 8 Operative schema of the "DPTto3D" component

for this contribution the segmentation that is currently in the alpha phase within the toolkit, the objective of monocular depth estimation is to determine the depth information of each pixel from a single RGB image without the use of any additional information. Through the use of the "DPTto3D" component is possible resorting to this technology to make a leap forward in terms of the amount of information that can be extracted from a simple 2D image in RGB colour scale (see Fig. 8).

Intel researchers have shared on their GitHub [20] page the main pre-training datasets (weights models) named MiDaS v3.1, v3.0 and 2.1. The MiDaS (Mixing Datasets for Zero-shot Cross-dataset Transfer) is a machine learning model [21] that has improved the compatibility of different datasets used to train neural models. The main differences in the depth information contained in the datasets are mainly a consequence of multiple detection tools used during the storage of information such as stereo cameras, laser scanners, and light sensors. For the right use of the "DPT-to3D" component, it is essential to download one of the weights models (which have a size between 80 MB and 1.5 GB) and to install the necessary Python libraries from the "requirements.txt" file as previously mentioned. Among these libraries, the *atoolkitdpt*[11] one was created by the author to facilitate the execution of this DPT technology inside the Grasshopper platform. From the diagram in Fig. 8, two operating modes can be deduced. The first concerns the return of *the depth map estimation* by delegating the user/designer to choose the source image and one of the 8 weights models previously locally downloaded—it is possible to customize the colour palette with Matlabplot library features. The second operation is more complex but is substantially linked to the generation of *the depth map estimation* using two different datasets with reduced size and located in the cloud: NYU Gen

[11] PyPI repository link, https://pypi.org/project/atoolkitdpt/, last accessed 2023/02/10.

v2 and KITTI.[12] Without digressing into technicalities, the codified tool showed greater coherence in the generation of three-dimensional outputs with the use of the NYU model. The latter has been developed by Nathan Silberman and his research team at New York University and specifically for the generation of depth maps for interior images. In fact, the dataset was trained in the fine-tuning phase with RGBD-scaled interior images captured by tools such as Kinect [22]. For this reason, when the designer activates the 3D mode one of the two datasets mentioned above will be used at the user's choice. The "DPTto3D" component can then show a preview of the final output of the generated point cloud or of a mesh object. However, the generated mesh object is not immediately usable, for this reason, as shown in Fig. 8, it is convenient to refine the point cloud using additional software such as Rhinoceros, Grasshopper or MeshLab. The DPT process stores in the ".ply" file format, all the geometric information related to the estimated point cloud and let the designer use it for further manipulations. Finally, a mention goes to the semantic segmentation process executable using the "DPTtoSS" tool actually in development (this tool will not be the subject of this discussion). The procedure is almost similar to that of generating the depth map estimation with weights models MiDaS, but in this case, the reference datasets are the *dpt hybrid-ade20k* and *dpt large-ade20k*, still provided both by René Ranftl's Intel research[13] group.

4 Toolkit in Action: Use Cases that Could Become Common in LTGMs-Driven Architectural Design

The use of AI-driven tools is expected to grow in the coming years as highlighted in the Stanford University Artificial Intelligence Index Report 2022 [23]. The report underlines that since 2018 the cost to train an image classification system has decreased by 63.6%, while training times have improved by 94.4% this means the AI becomes more affordable and higher performing as the lower costs are equivalent to a prelude of a wider deployment of such products. Indeed, as vast and seemingly unknown as it may seem, the next logical step is the understanding of how such tools could be adopted by the AEC industry. The design phases in the architecture, design and engineering sectors are multiple however, it is quite usual to recognize in the actual design phase three moments completely dedicated to design development and entirely pervaded by functional and emotional creativity, namely: Pre-Design (PD), Schematic Design (SD) and Design Development (DD). Generally, the PD phase can last from 2–4 weeks, the SD phase from 4–8 weeks and the DD phase from 8–12 weeks. The tools developed and described so far have the potential value of settling between the final Pre-Design phase and totally in the Schematic Design

[12] Global–Local Path Networks for Monocular Depth Estimation with Vertical CutDepth, https://github.com/vinvino02/GLPDepth, last accessed 2023/02/10.

[13] DPT weights models for the semantic segmentation, https://github.com/isl-org/DPT, last accessed 2023/02/10.

phase even being able to provide graphic support in the Design Development phase. The Schematic Design stage is the time for imaginative thinking, this phase enables architects and designers to experiment with various design concepts and gives them the opportunity to experiment and conceptualize the overall aesthetic appeal. On the other hand, the expectations for contemporary design presentations call for a seamless blend of photorealism and dynamic multi-dimensionality, enabling clients or buyers to make informed decisions. The image creation process is a challenging and time-consuming endeavour, requiring meticulous attention to detail and a keen sense of creativity. Fortunately, the advent of the LTGMs technologies has alleviated these pressures, delivering stunning visuals at lightning speed. The topics in which a designer is generally involved in the SD phase can be enclosed in a simple concept: the production of conceptual images that are inspiring, and emotional that offer opportunities for observations in investigating the design act. Therefore, the basic scheme of use of ATk is proposed below by breaking down the concept described above into the three operative phases: Creation, Variation and Editing (Fig. 9).

Asking to GPT-3

The designer can enter the prompt directly or decide to query the neural model through the "OpenAI-GH_Ask" component. This can preclude the automatic generation of a text that can be used directly in DALL-E or Stable Diffusion, or it can help the designer in formulating the descriptive text with better words. Following three examples:

– request for the generation of a text-prompt;

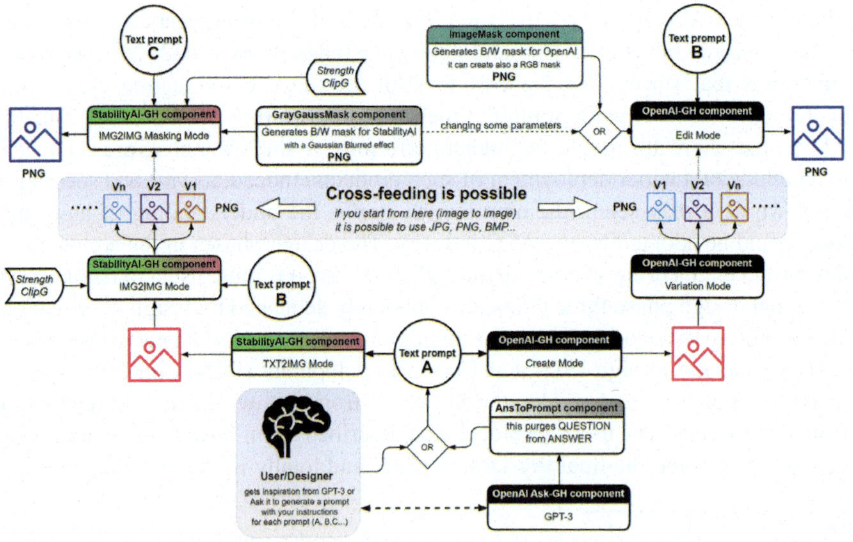

Fig. 9 Operative schema of the Toolkit usability

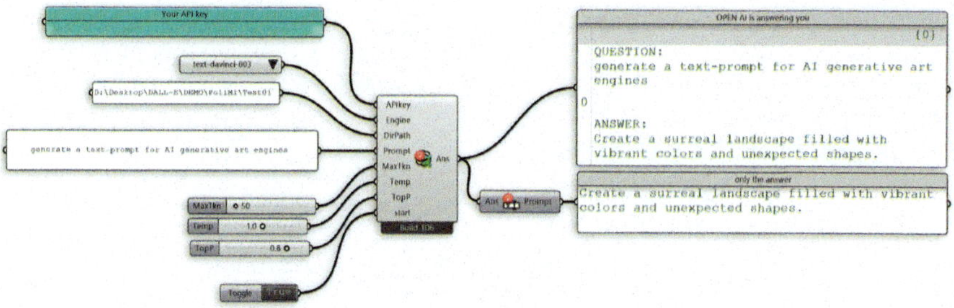

Fig. 10 Generic text-prompt generated by OpenAI GPT-3 model

- suggestion/alteration of a text-prompt assigned by the designer with the relative generation of the image model;
- suggestion of response to a specific request (e.g.: focused on environmental assessment) (Figs. 10, 11 and 12);

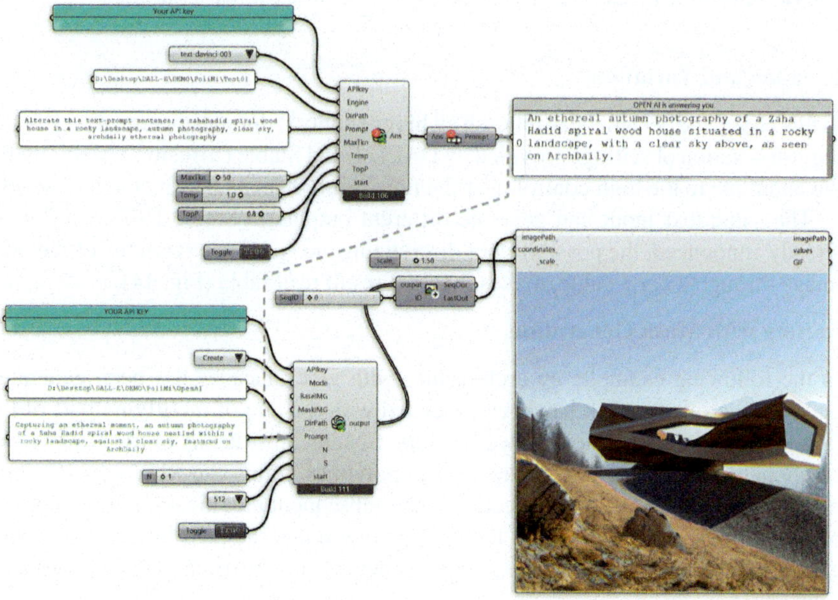

Fig. 11 (Top) Alternative way to write a designer's text-prompt; (bottom) DALL-E text-to-image generation from the alternative prompt

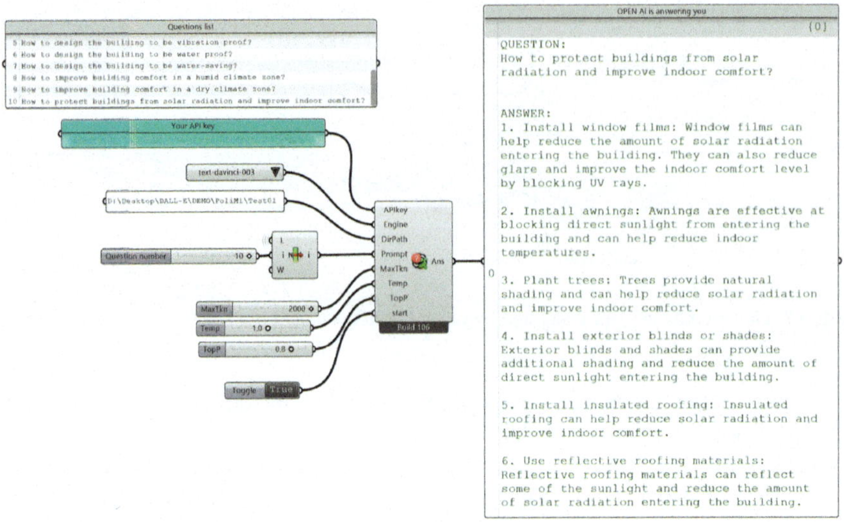

Fig. 12 OpenAI GPT-3 model answer to a question focused on environmental design

Creation and Variation

In this phase, the designer can unleash all his creativity[14] through the use of words and the combination of settings provided by DALL-E and Stable Diffusion, especially for the latter due to the high control over the final result that can be generated (Fig. 13).

The same text input can generate different outputs in Stable Diffusion, but as already announced, the possibility of discretizing the "parameterization" of the final image through a *Seed* value, makes the final result restorable (Fig. 14).

Editing with Mask Generation

In the following example, an architecture with green facades has been generated on which an image-mask has been modelled with the aim of modifying part of the facade according to the second text prompt. Specifically, thanks to the "ImgMask" component, it was possible to draw some polylines within the Rhinoceros[15] environment delimiting the areas affected by the subsequent modification through a text prompt that included the word "loggia". At this stage, the possibilities are almost endless although DALL-E does not allow for detailed control such as Stable Diffusion (Fig. 15).

In the Fig. 16 the designer needs to pay attention in generating the image-mask because of the quality of the edited image depends on how he has mixed the greyscale or the black/white colours.

[14] A further example regarding the variation is provided by the author in this video demo, https://youtu.be/UPTc5o6rOLM, last accessed 2023/02/13.

[15] A further example regarding the editing is provided by the author in this video demo, https://youtu.be/nmJTP1QvuIE, last accessed 2023/02/13.

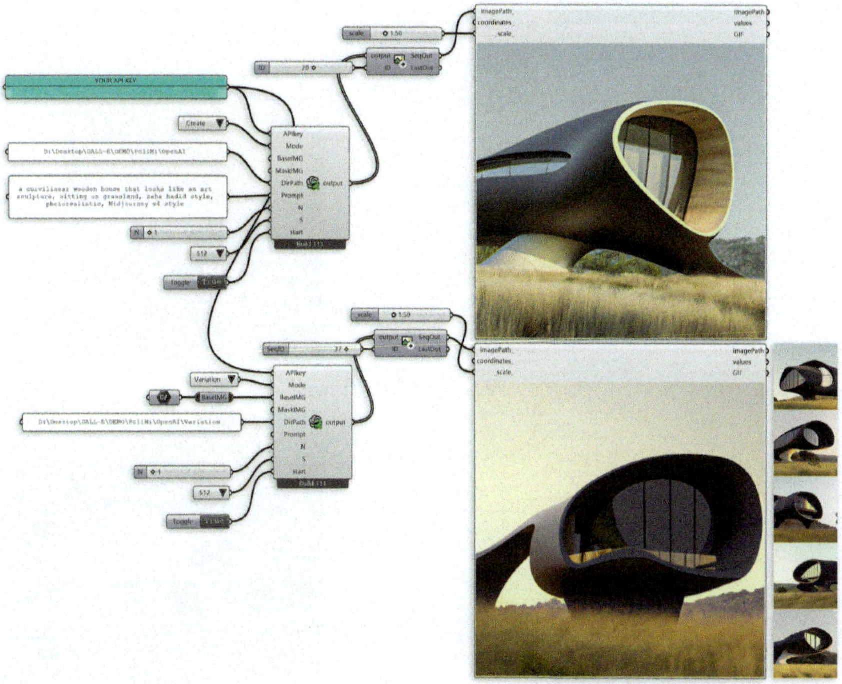

Fig. 13 DALL-E text-to image and image-to image processes; at the bottom different variations of the *BaseIMG* (top image)

What has been discussed up to now concerns the basic actions shared by both platforms, however, through the Toolkit, the designer will be able to mix the use of these tools passing from one neural model to another, benefiting to the maximum from their respective strengths.

4.1 Generative Models AI-Powered as a Potential Rendering Engine

One of the noteworthy potentialities of LTGMs is that of exploiting the ability to act on the image-mask and on a well-considered textual input to transform the neural model into a potential rendering engine without running into technical issues or time-consuming architectural visualizations. The proposed experiment starts from an extremely simplified 3D interior design model created with the NURBS Rhinoceros modeller (Fig. 17).

Clearly, this potential currently requires further investigations in order to improve the coherence of the final image with the source one. In this sense, in Fig. 18,

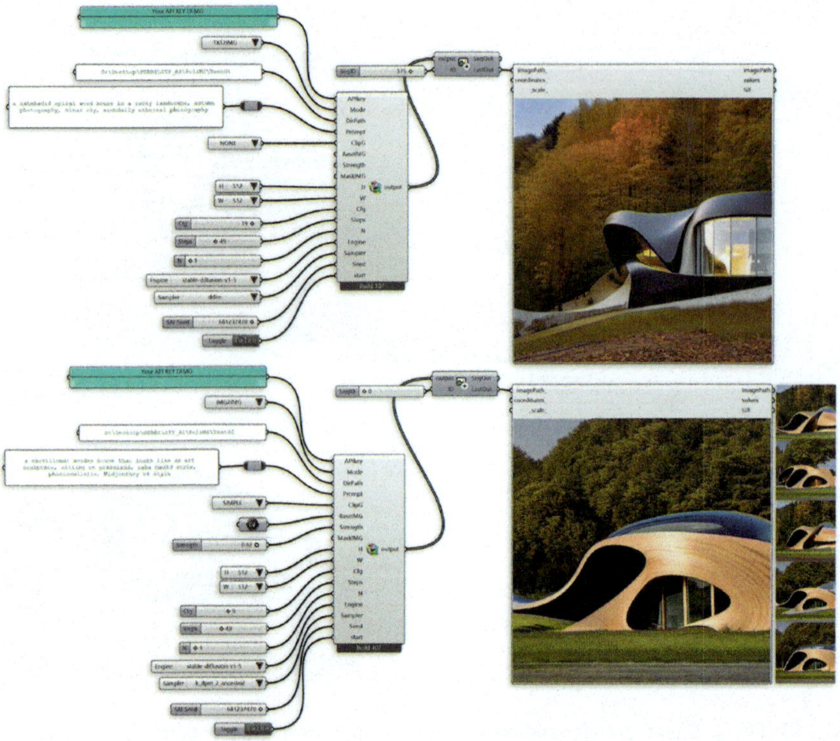

Fig. 14 Stable diffusion text-to image and image-to image processes; (bottom) a variation of the *BaseIMG* (top image) driven by second text prompt, *Strength* and *ClipG* parameters

through the application of the depth map estimation, it is possible to generate a grayscale image (ATk Gray palette) of the 3D model and obtain a more coherent representation, but not detached from some "poetic license" of the AI model.

4.2 What You See Is What You Text (WYSIWYT): From Monocular Depth Map Estimation to 3D Representation

In 1970 at the Xerox Palo Alto Research Center, the Hungarian-American computer researcher and astronaut Charles Simonyi coined the expression "What you see is what you get", referring in particular to the possibility of obtaining outputs identical to the way they were conceived—particularly well-known expression in the use of HTML code for websites. Here, however, the provocative expression "What you see is what you text" is used as a demarcation line for future AI technologies capable of rendering, in three dimensions, what has been described textually through LTGMs and DPT technologies. The experiment conducted is operationally described

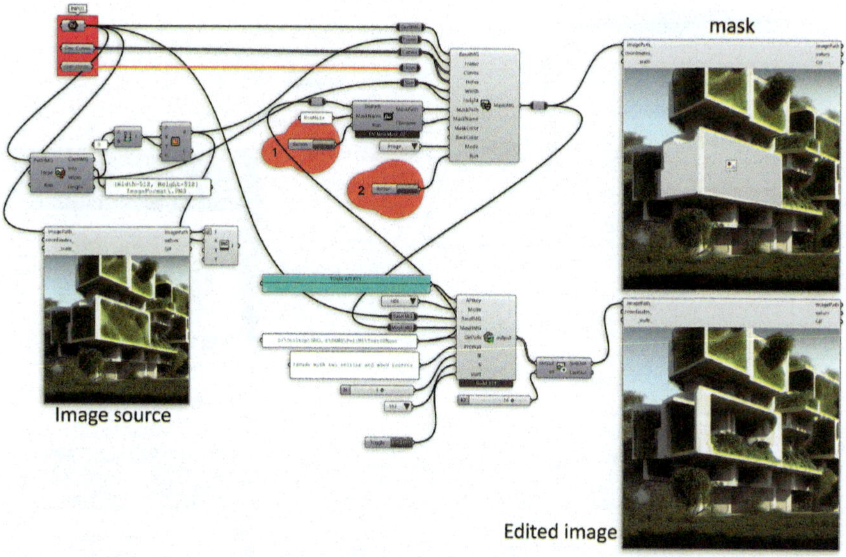

Fig. 15 DALL-E image-to image processes using an image-mask; (top right) polylines as curves selection have defined the image-mask; (bottom right) edited image driven by the second text prompt;

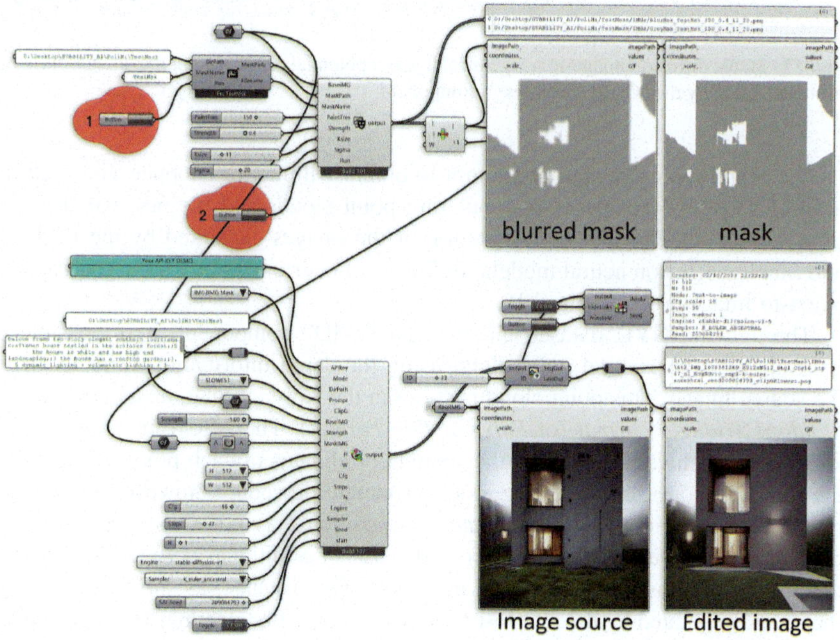

Fig. 16 Stable diffusion image-to image processes using a gaussian image-mask; (bottom) the edited image driven by the second text prompt, *Strength* and *ClipG* parameters

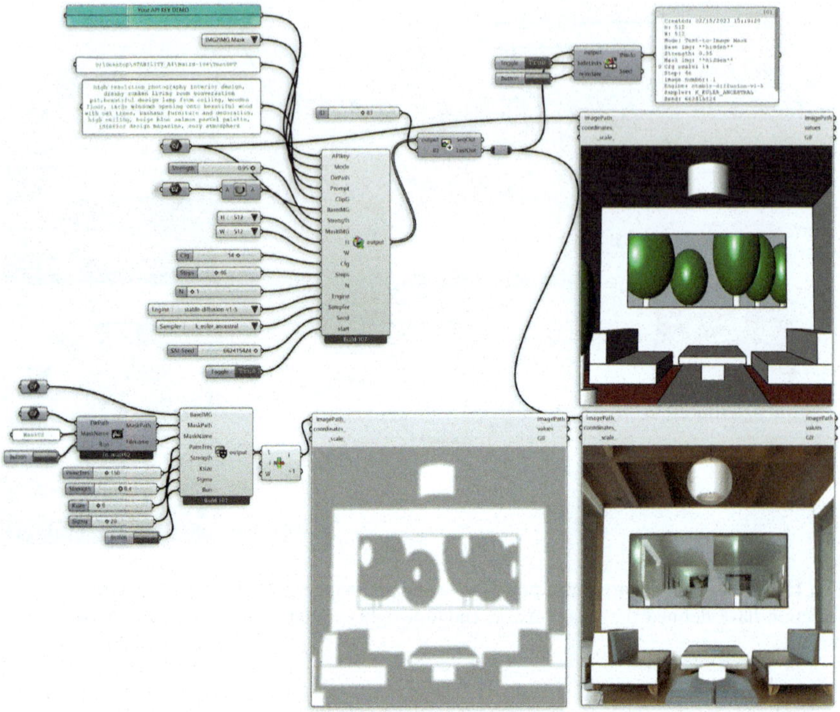

Fig. 17 Stable diffusion image-to image masking as a potential rendering engine, image-mask has been generated by the "GrayGaussMask" component

in Fig. 8 and represents a first attempt to obtain within Grasshopper, and with the help of software capable of manipulating point clouds and meshes, when necessary, a three-dimensional representation of the images generated by the DALL-E and Stable Diffusion neural models. Below is an interior design image generated as a text-to-image process (Fig. 19).

Through the NYU dataset and the "DPTto3D" component, it is possible to generate an experimental point cloud using the depth information extracted from the source image. This point cloud is saved in the ".ply" format, also known as the *Stanford Triangle Format* (or *Polygon File Format*). The ASCII encoding of this kind of file archives a list of all the spatial coordinates of each point belonging to the aforementioned cloud. Being an experimental procedure, however, the images produced and their three-dimensional transposition may not always lead to well-defined mesh objects, for this reason, the use of additional software or specific definitions of Grasshopper may be necessary. For instance, as shown in Fig. 20, the image generated by the text-to-image process was first decimated and improved through the MeshLab software and then imported into Rhinoceros to be displayed as a "boxed-style" for pure purpose illustration or as a ".obj" mesh object. The final three-dimensional object is clearly susceptible to improvements since it has

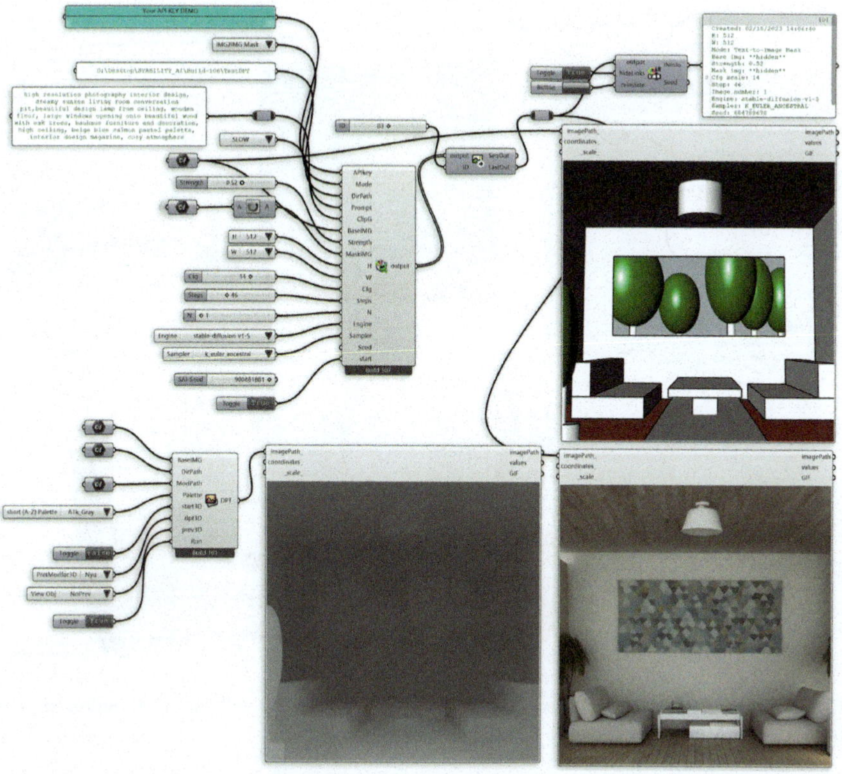

Fig. 18 Stable diffusion image-to image masking as a potential rendering engine, image-mask has been generated by the "DPTto3D" component

not been directly detected as happens for the laser scanner survey, nonetheless, the motto reported at the beginning with the acronym WYSIWYT[16] eloquently shows the reader that what has been textually described in input it is exactly what he sees three-dimensionally within the modelling software used.

5 Conclusions

The advent of artificial intelligence in the consumer world is destined to alter and modify products and productions of any category and this is due to the constant reduction of the training costs of the neural models while the hardware enhancement reduces the computation times. The open-source policy, or at least code sharing, is accelerating the adoption of these new technologies. These two aspects, the reduction of sustainable costs and speed of adoption translate into a single concept of the

[16] Video demo, https://bit.ly/LA-DPTto3Dobj, last accessed 2023/02/21.

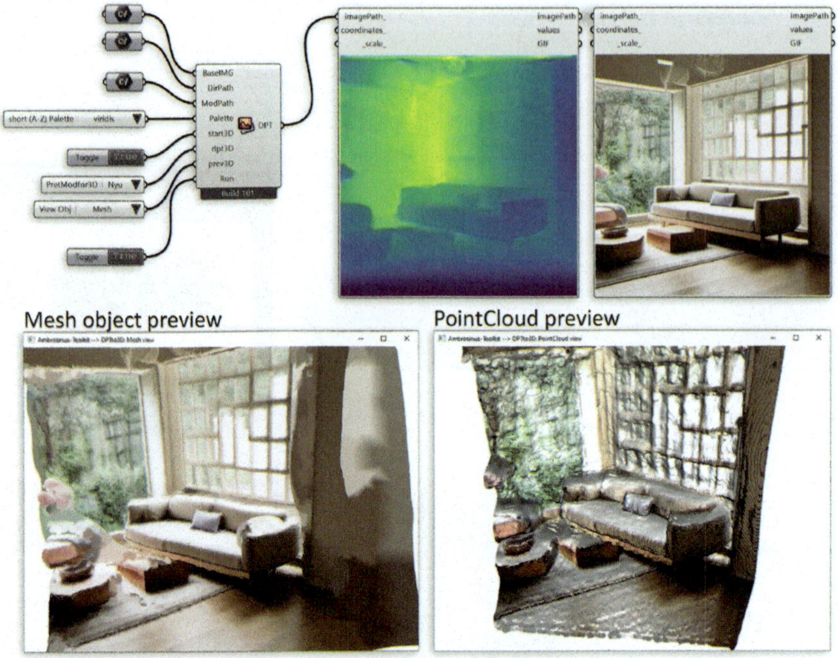

Fig. 19 (Top) "DPTto3D" component generated the depth map estimation; (bottom) preview frames of the Mesh object and the PointCloud prediction

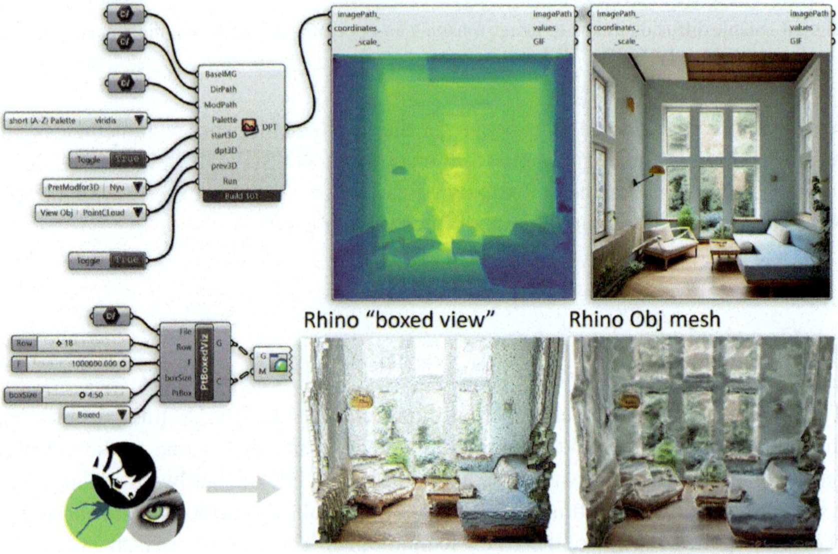

Fig. 20 (Top) "DPTto3D" component generated the depth map estimation; (bottom) preview inside Rhinoceros after some actions with and without specific software

scalability of innovation. Georgetown University's Think-Tank, the CSET Center for Security and Emerging Technology [24], has ascertained that the "AI war" is mainly taking place between the USA and China, however, if they consider the whole scientific production not only disseminated in English but also in Chinese, the gap becomes far greater than 5 times over the USA—the warning for other nations is eloquent to avoid significant technological gaps. Once again, the driving force behind the innovation still remains the scientific research and the ability of some industrial sectors to let themselves be cross contaminated by all technological advances, that of architecture and engineering is a valid test bench.

The AEC industry is already showing the symptoms of the fifth industrial revolution of the advent of a new paradigm shift in the field of architecture and engineering, architects such as Makoto sei Watanabe and Wanyu He founder of XKool Technology, have coined new terms such as "architectural intelligence", "AItect" and "AI-chitect" emphasizing the vector direction of modern design firms and the leap forward of computational design. With technological progress evolving at these speeds, the AEC industry will be further invested by a strong cultural and operational dichotomy, in particular, that defined by consumers and developers, by those who will be conveyed by the collective technical culture dictated by the Big Companies and those who will carry such technologies and tools. However, the possibilities offered by the digital age can provide interesting prospects for those professional roles that can be defined as "devsumers" cultural-crasis of the two groups just mentioned—practically the concept of "prosumer" introduced in the 70s by Alvin Toffler and now transposed into the era of scripting and coding. These roles are able to actively take part in the design innovation challenges activities measurable through the noble art of developing their own tools with a certain autonomy and awareness. A well-known historical practice during the Renaissance age that the author defined Brunelleschi's paradigm [25]. The Toolkit developed, represents one of the possible approaches that can be adopted in professional practice, namely the possibility of integrating strategic aspects with technical ones to optimise and manage the design accomplishment. The new Digital Era through coding, computational thinking and the advent of AI will accustom the AEC insiders to increasingly deal with *metatools*, infrastructures dedicated to the development of tools and the decoding and re-using of different technologies. The tools developed to integrate LTGMs and DPTs to support the schematic design phase have provocatively in a scaling way addressed these issues.

In the future, the Toolkit will require further implementations and measures to enhance the user experience and its pertinence in the architects' workflow in everyday work [26]. For instance, the semantic segmentation, the possibility of running the various tools locally[17] and some advances in the potential use of AI as a rendering engine are currently in the alpha phase, implementations partially conveyed by the feedback received from the first beta testers.

[17] In the last few months Ambrosinus-Toolkit has received several implementations, including the possibility of using Stable Diffusion on a local machine and as a rendering engine, thanks to the project called "Automatic1111". ATk development link, https://bit.ly/SDandCNinsideGH, last accessed 2023/06/23.

Nonetheless, the digital message enclosed by the artisan practice of building tools as prerequisite assumptions that anticipate the design act aims to grasp the intrinsic aspects of industriousness and creativity resulting from human intelligence. The sociologist Richard Sennet in his book "The Craftsman" focused on the valuable meaning of tools underlines precisely the ability they have in stimulating the craftsman in solving problems, in exciting his mind and in preparing it for the so-called leap into the unknown. In the AEC industry and generally in technical work, the most common feeling that threatens the trust placed in the feeling of "possible", of "being able in doing", is that of frustration dictated by the limitations, albeit temporary, of a given tool or by its potentiality yet unspoken. Therefore, the action of the so-called *devsumers* can be traced back to a heuristic approach that brings the domain of the imagination closer to the idea of progress, or better to the acquisition of skills that aim to advance and improve user's experience and developer's one both as much as possible. Through the limitations of a tool or while waiting for its availability it is the inventiveness that makes up for the design needs, favouring the emergence of the sense of competence that transforms the tool development experience, and the coding practice, in a technical metaphor that stimulates the creative thought and its revealing capacity.

References

1. Domingos P (2015) The master algorithm: how the quest for the ultimate learning machine will remake our world. First paperback, Basic Books, New York
2. Article link. https://bit.ly/SneakPeek-Expo2021-iMesh. Accessed 1 Feb 2023
3. Chaillou S (2019) The advent of architectural AI. Towards Data Science. Article link. https://towardsdatascience.com/the-advent-of-architectural-ai-706046960140. Accessed 2 Feb 2023
4. Chaillou S (2020) ArchiGAN: artificial intelligence x architecture. In: Architectural intelligence selected papers from the 1st international conference on computational design and robotic fabrication (CDRF 2019). Springer, Singapore, pp 117–127
5. Sevilla J, Heim L, Ho A, Besiroglu T, Hobbhahn M, Villalobos P (2022) Compute trends across three eras of machine learning. In: International joint conference on neural networks (IJCNN), pp 1–8. https://doi.org/10.1109/IJCNN55064.2022.9891914. Accessed 2 Feb 2023
6. A high resolution of the chart entitled "three eras of the ML" elaborated by the author, is available here. https://bit.ly/LA-3ErasMLchrt. Accessed 7 Feb 2023
7. The State of AI Report analyses the most interesting developments in AI. The Report is produced by AI investors Nathan Benaich and Ian Hogarth. https://www.stateof.ai/. Accessed 2 Feb 2023
8. "Law of Accelerating Returns" article link. https://www.kurzweilai.net/the-law-of-accelerating-returns. Accessed 2 Feb 2023
9. Article link. https://openai.com/blog/introducing-openai/. Accessed 3 Feb 2023
10. Hy L (2022) Large-scale artificial intelligence models. Computer 55(05):76–80
11. AI-text classifier homepage. https://platform.openai.com/ai-text-classifier. Accessed 3 Feb 2023
12. Techcrunch article. https://tinyurl.com/NewYorkPublicSchool. Accessed 3 Feb 2023
13. Article link. https://tinyurl.com/MITtechReviewArticle. Accessed 4 Feb 2023
14. Article link. https://blog.google/technology/ai/bard-google-ai-search-updates/. Accessed 7 Feb 2023
15. Ho J, Jain A, Abbeel P (2020) Denoising diffusion probabilistic models. ArXiv abs/2006.11239. Accessed 3 Feb 2023

16. Nichol A, Dhariwal P, Ramesh A, Shyam P, Mishkin P, McGrew B, Sutskever I, Chen M (2021) GLIDE: towards photorealistic image generation and editing with text-guided diffusion models. arXiv preprint arXiv:2112.10741. Accessed 2 Feb 2023
17. Ramesh A, Dhariwal P, Nichol A, Chu C, Chen M (2022) Hierarchical text-conditional image generation with CLIP Latents. ArXiv abs/2204.06125. Accessed 2 Feb 2023
18. Inpainting and outpainting prototype tool homepage. https://github.com/lkwq007/stablediffusion-infinity. Accessed 7 Feb 2023
19. Rombach R, Blattmann A, Lorenz D, Esser P, Ommer B (2021) High-resolution image synthesis with latent diffusion models. In: Proceedings of the IEEE/CVF conference on computer vision and pattern recognition (CVPR), pp 10674–10685
20. Intel Isl work, and models are freely. https://github.com/intel-isl/MiDaS. Accessed 9 Feb 2023
21. Ranftl R, Lasinger K, Hafner D, Schindler K, Koltun V (2019) Towards robust monocular depth estimation: mixing datasets for zero-shot cross-dataset transfer. In: Proceedings of the IEEE transactions on pattern analysis and machine intelligence, vol 44, pp 1623–1637
22. Silberman N, Hoiem D, Kohli P, Fergus R (2012) Indoor segmentation and support inference from RGBD images. Lecture notes in computer science, vol 7576. Springer, Berlin, Heidelberg, pp 746–760
23. Stanford University Human-Centered AI report. https://aiindex.stanford.edu/report/. Accessed 10 Feb 2023
24. CSET full report. https://cset.georgetown.edu/publication/counting-ai-research/. Accessed 16 Feb 2023
25. Ambrosini L (2018) Data, digital & design—Produzione del progetto digitale e processi decisionali: la progettazione "flessibile" nell'Era dello Scripting e del Building Information Modelling come nuovo paradigma tecnologico. PhD thesis in Architecture XXXI cycle, DiARC, University of Naples "Federico II". https://doi.org/10.13140/RG.2.2.27158.29769
26. Turchi T, Carta S, Ambrosini L, Malizia A (2023) Human-AI co-creation: evaluating the impact of large-scale text-to-image generative models on the creative process. In: Spano LD, Schmidt A, Santoro C, Stumpf S (eds) End-user development. IS-EUD 2023. Lecture notes in computer science, vol 13917. Springer, Cham. https://doi.org/10.1007/978-3-031-34433-6_3

Luciano Ambrosini is an architect with a Ph.D. in architecture from the University of Naples "Federico II". He is an independent researcher focused on environmental and computational design. He has published scientific articles on parametric design, BIM and the computational approach to architecture. He has consulted for the Apple Developer Academy of Naples, worked with i-Mesh Company (Italy) for the Expo Dubai 2020 and Foster & Partners (UK) as an environmental design analyst. He is attentive to the integration of new technologies even AI-powered to enhance architectural design with the development of tools and workflows based on team expertise.

Chapter 2
EOC ECO_2 Plugin: Automated Embodied Carbon Calculations Within Autodesk Revit

Duncan Walters and Chiara Bariviera

Abstract The embodied carbon (EC) emissions of the construction industry are governed by the local carbon intensity of energy production used in raw material extraction, product manufacturing & fabrication, transportation, and erection of buildings. Eckersley O'Callaghan (EOC) has created EOC ECO_2, an automated add-on for Autodesk Revit, allowing the building sector to record emissions directly from design models. Recent projects were compared with targets set by the Royal Institute of British Architects (RIBA) and the London Energy Transformation Initiative (LETI), with the majority meeting current guidelines. Primary performance levers included structural grid size, material choice and the extent of permitted demolition. The UK construction industry can achieve target guidelines for new construction until around 2030, after which a more targeted approach will be necessary using circular economy principles, more refurbishment projects and increased use of regenerative materials. Data extracted by Building Information Modeling (BIM) should allow for ease of knowledge share across the industry. A similar approach to monitoring of EC in infrastructure is recommended.

Keywords Emissions · Carbon · Materials · Autodesk · BIM

1 Introduction

The discourse surrounding the decarbonization of the UK power grid is already in progress, highlighting the advantages of clean energy as a positive step towards reducing emissions. However, the construction industry has fallen behind, still

D. Walters (✉) · C. Bariviera
Structural & Civil Engineering Department, Eckersley O'Callaghan, 236 Grays Inn Road, London WC1X 8HB, UK
e-mail: duncan@eocengineers.com

C. Bariviera
e-mail: chiara@eocengineers.com

P. Ruttico (ed.), *Coding Architecture*, Digital Innovations in Architecture, Engineering and Construction, https://doi.org/10.1007/978-3-031-47913-7_2

heavily reliant on carbon-intensive practices in raw material extraction and manufacturing processes.

"Embodied carbon refers to the greenhouse gas emissions arising from the manufacturing, transportation, installation, maintenance, and disposal of building materials".

According to the World Green Building Council, building and construction activities account for 39% of energy-related carbon dioxide (CO_2) emissions. Taking this into account alongside the Paris Agreement which aims to limit global warming to 1.5'C by 2050, it is vital that we dramatically drive down CO_2 emittance within the construction sector.

The first hurdle for Eckersley O'Callaghans Structural team, in its mission to reduce project emissions, was to understand how to measure embodied upfront carbon. As structural engineers our impact can be as much as 65% of the entire embodied carbon spent at the outset of the project during supply and construction activities.

The Institution of Structural Engineers have clearly set out the method in 'How to calculate embodied carbon, 2nd Edition' (HTCEC) for recording and a few corporations have committed to CO_2 reduction.

This report focuses only on the structural embodied carbon encountered during the **Product Stage (A1-A3)** and the **Construction Stage (A4-A5)**. There are benefits to consideration beyond the Construction Stage however the structural influence is negligible until the "End of life stage" and further "Circular economy" approaches may be implemented.

2 Targets

Following the 2021 alignment of UKGBC, LETI and RIBA EC targets, the below Fig. 1 indicates the proposed overall building typology targets and a breakdown for each respective discipline and project stage for the upfront carbon budget (Fig. 2).

3 Identifying the Need for the Software.

Not enough reliable data exists for the construction industry to reflect on its impact and in turn determine where opportunities and improvements exist. Also, it is unlikely to be able to self-regulate without external pressures.

It is difficult to plot the route towards NetZero2050 without understanding our current and historic baseline score, i.e. if we don't have a view of our current product as relates EC then we do not have adequate perspective to improve or consider ways in which improvements can be made.

Retrospective analysis of recent work will allow us to determine the main drivers for change in the industry. An automated approach would make this task far easier.

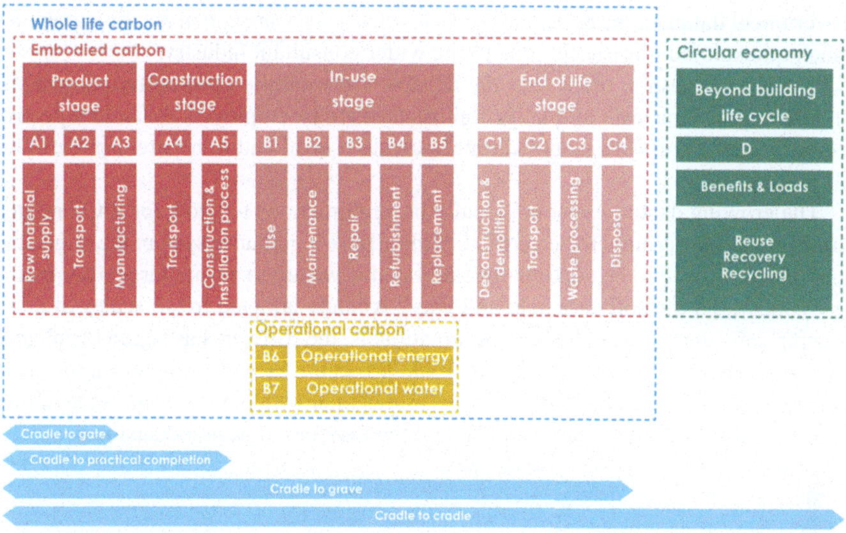

Fig. 1 System Boundary: EN15978:2011 Display of modular information for the different stages of the building assessment. RICS November 2017 (Whole life carbon assessment for the built environment, 1st Edition)

2020 LETI Targets (kgCO2e/m2)			
Building Use	Total Embodied Carbon Target (Architecture MEP and Structure)	Structure Only (A1-A3)	Structure Only (A1-A5)
Office	600	353	390
Education	600	260	276
Small Residential	500	265	285
Medium to Large Residential	500	290	335

2030 LETI Targets (kgCO2e/m2)			
Building Use	Total Embodied Carbon Target (Architecture MEP and Structure)	Structure Only (A1-A3)	Structure Only (A1-A5)
Office	350	206	228
Education	350	152	161
Small Residential	300	159	171
Medium to Large Residential	300	174	201

Fig. 2 Total and Structural specific extrapolated EC targets based on LETI Guidance for best practice in 2020 and 2030. Eckersley O'Callaghan 2019

In late 2022 the UK's Net Zero Carbon Buildings Standard (NZCBS) has requested a call for evidence to populate the Built Environment Carbon Database (BECD) with recorded industry project scores. Whilst EOC are uploading project scores onto this government database we hope that the EOC ECO_2 software can make this stock take much quicker and more accessible to the wider consultant industry.

There is a question mark over the accuracy of multiple third-party assessments and therefore some level of quality control needs to be employed. As a retrospective assessment the software would certainly create a level and transparent process for all projects.

The growing databases should ideally be agglomerated to allow for IPCC related limits to global warming to be verified. Similarly, to a guidance system the target and the trajectory need to be continually calibrated if we are to ensure targets are met.

Who should the software help and why? Already adopted by all scales of engineering consultancy, from SME to multinationals, the software has begun to enhance everything from domestic to the largescale projects. As well as other engineering and architectural disciplines who wish to assess their own outcomes and Regulatory bodies, whom are both keen to understand the nuances of design decisions.

4 The Plug-In

Using the RICs 2019 ICE database and applying the Global Warming Potential coefficients to each building volume is a simple calculation but you need to have all the material volumes to do this, which until now has been a painstaking manual exercise.

One of the main drivers for the software development which exists on a BIM platform was to ensure that input from BIM staff could be minimized, allowing engineers to lead the process with an easy-to-use interface. The software should be versatile, easily downloaded and deployed, and should comprise traceable data with black-box features removed so that the data can be investigated—a tool made by engineers for engineers.

It has the potential to be used across other industries which use the Autodesk Revit platform and similarly the code itself can be borrowed for use on other platforms—something we have already begun to progress with in parallel.

EOC ECO_2 opens as a pop-up from the add-ins ribbon (Fig. 3) as a user interface adjacent to the default 3d view of Autodesk Revit giving a 3D 'Digital Twin' of the building design and a visual breakdown of the EC for all of the various structural typology elements modelled. The data can be manipulated within the plug-in without having to remodel things in Revit; materials can be added, changed or re-allocated and re-factored without having to change the BIM model itself.

As soon as the pop-up window is opened, the application programming interface (API) the reader code, filters and extracts the volumes, collects their material names and classes and filters all of those elements into a few basic structural material groups. Any errors resulting from the automated sort and filter can be corrected by adjusting the populated table on the pop-up window. If a material property has been incorrectly identified the parent material and specification can be adapted manually without changing the Revit model itself.

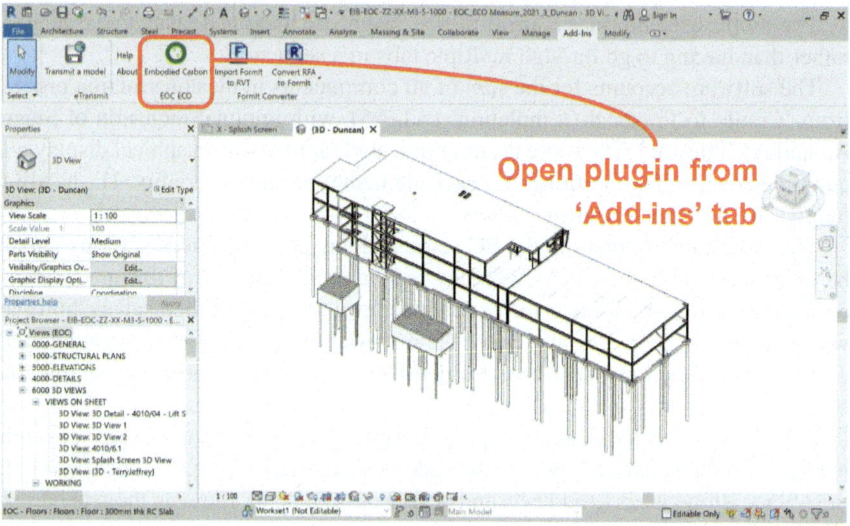

Fig. 3 Launching plugin from revit. Eckersley O'Callaghan 2019

Each standardized material has pre-loaded global warming potential (GWP) values within the script and therefore it provides a best guess fit for each element. A simple task of 'sense-checking' these outputs to allow for corrections to be made (Fig. 4).

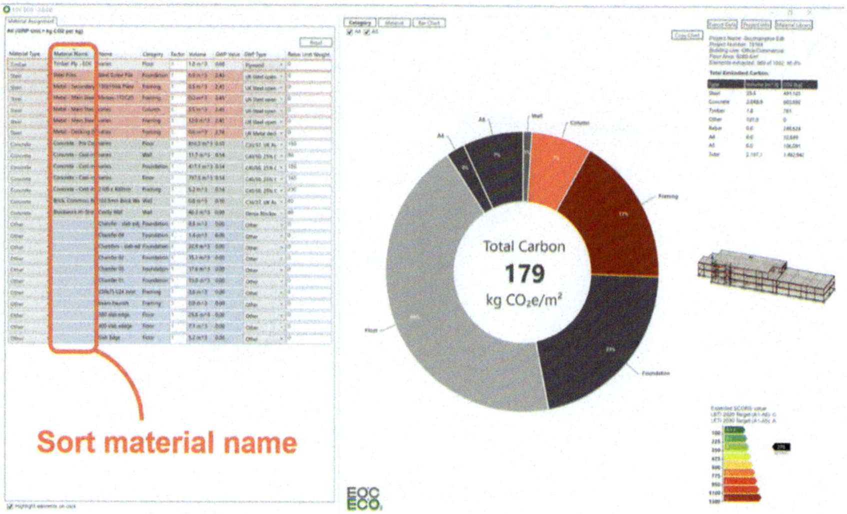

Fig. 4 View of pop-up window EOC ECO_2 software plugin

The simple user interface is an all-in-one dashboard making it easier to see changes rather than having to go through multiple tabs in a sequence.

The software accounts for the sum of all components in the construction process from 'Cradle-to-Practical-Completion' (A1-A5), with optional inclusion of Stages A4 and A5 if desired. The score is communicated rapidly with graphical display and quantified data and proportioned for each element type and materiality. The headline outputs are immediately clear.

The largest contributors to the EC score and any areas that need scrutiny should be evident from the outset. As with any software, the scrutiny of this analysis and output data is only as good as the experience level of engineer themselves, however with a relatively short learning curve the reviewal process becomes second nature. The pie chart function also allows for distillation of element category and material type.

A basic set of material specifications, covering timber, steel and concrete, is preloaded onto the software however the 'add new materials' (Fig. 6) function within the library allows users to add new and bespoke coefficients allowing them to be very specific about their projects.

We can see from the below case study that the steel framing accounted for 40% of an entire score, whilst the precast concrete floor panels only contributed 27%, and columns, foundations and walls (~23%) made up the remaining material constituents. The transportation of materials only accounted for 1% and site construction phase recorded at 8% of the total budget. It is common for A1-A3 to account for 80–90% whilst A4-A5 to account for 10–20% (Fig. 7).

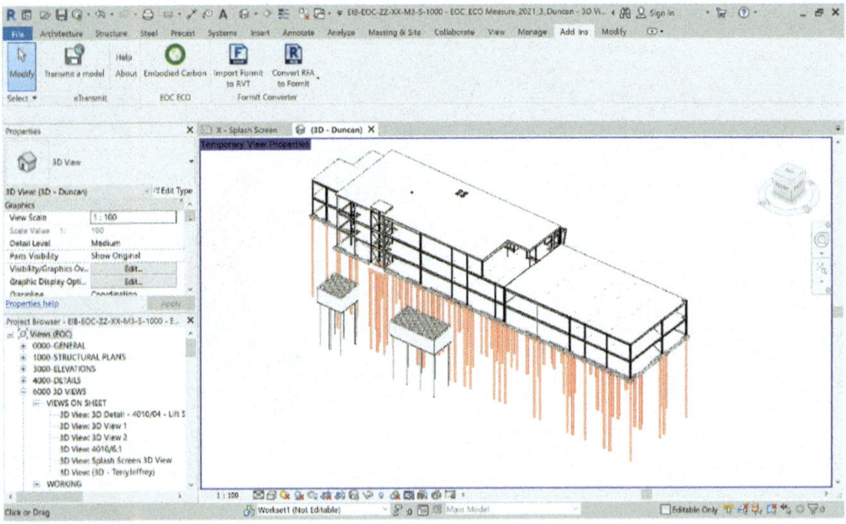

Fig. 5 View of highlighted piled foundation elements within the model during a 'sense-check' of the initial sum—materials identified need to be checked against the design specification

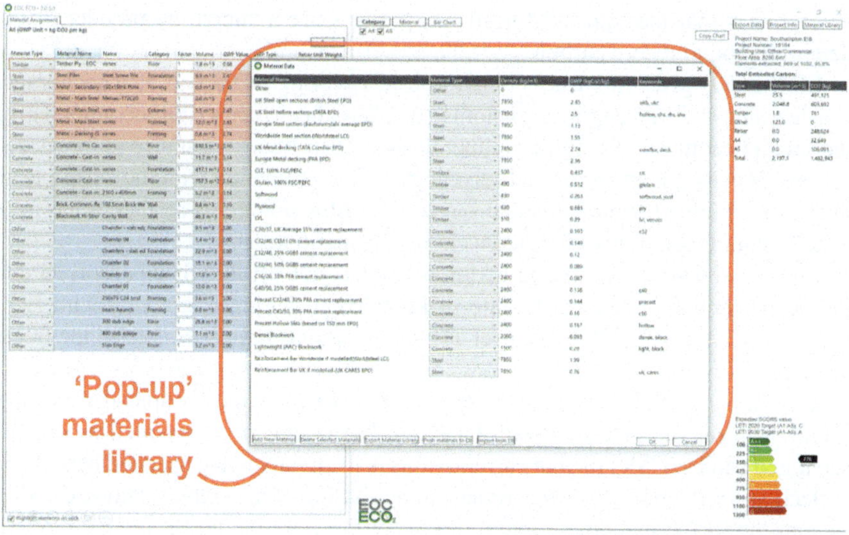

Fig. 6 View of basic materials library with 'add new materials' functionality

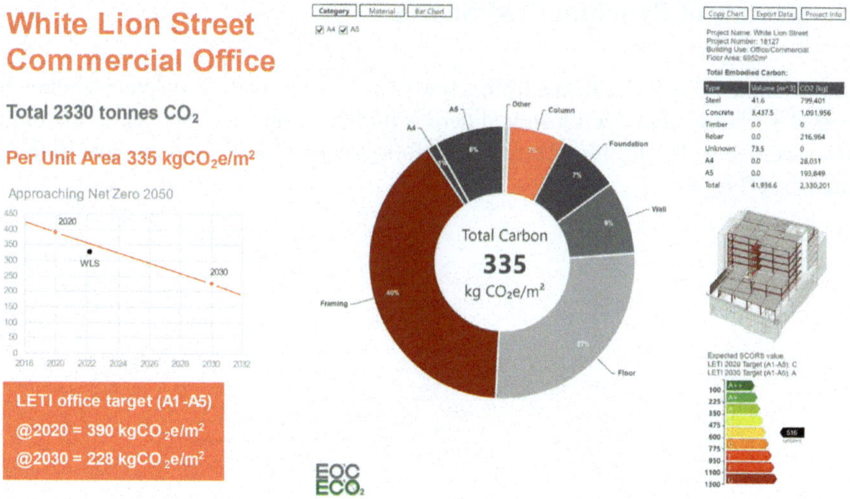

Fig. 7 Example Project Structural Embodied Carbon Score, with key project data and related LETI Target curve plotted for Commercial Office Usage

How Has the Software Helped to Improve Data Retrieval?

Prior to the climate declaration, the only incentive for engineers to measure tonnages was to explore the commercial benefits for our clients, but since the metric of upfront carbon relies on accurate volume calculation, and one hugely beneficial outcome for surveyors is that the 'quants' for all our projects are a simple by-product of our determination to record carbon.

EOC ECO_2 gives full access to hard project scoring metrics. It also encourages self-learning amongst engineers, whom to date have been focused on accommodating the 'main headlines' of their discipline; those criteria being resistance to forces and actions such as gravity, wind and seismicity and coordination with the design team to ensure building functionality and services coordination. Even aesthetics has been a more natural perspective for engineers to consider until now.

Embodied Carbon is a recent addition that has not yet fully been absorbed into the design process however the rate of knowledge sharing is growing with EOC ECO_2 playing its part. Globally, over 150 organizations have downloaded the freely available software.

Unfortunately, the software does not have the ability to remotely upload the project scores to a remote or centralized database. The data can be exported as comma separated values (CSV) file but currently this is local only to the user and hence data collection for the wider industry relies on the sharing of data via to statutory bodies such as the BECD and other groups.

5 Review of Benchmarked Scores

An example of early 2020 data from a few EOC project scores is displayed below in Fig. 4.4. The sample size illustrated here is limited in number and clearly the more data we can gather from our project portfolio the more faith we can have in these statistics (Fig. 8).

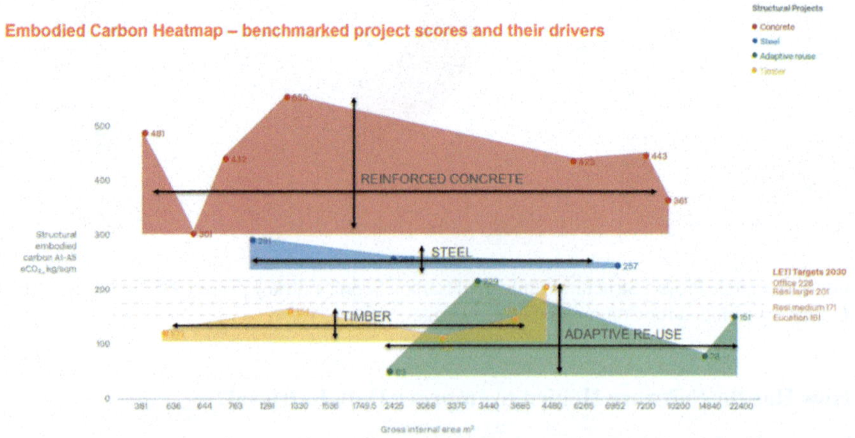

Fig. 8 Retrospective and current project benchmarks scores

However, immediately we can see some strong correlations between primary structural materials and the A1-A5 scores of our dataset. The primary structure most frequently scoring under the 2030 LETI best practice targets are those using renewable materials such as timber as a primary structure. Our more traditional framing materials of concrete and steel apparently have a higher EC emittance generally.

The other noticeably successful group was the 'adaptive re-use' sector—those projects involving significantly large parts of structure retained from the previous development. It stands to reason that projects avoiding full demolition should score well given that their premise is to limit material waste. Less new materials equate to lower carbon intensity.

Also, by inspection the gross internal area and scale of the building appears to have a little impact on the overall material advantages of timber and refurbishment approach.

It is not identified in the illustration but worth noting that projects involving basement excavation naturally carry a greater carbon budget related to their high material volume demand. They must accommodate high earth and hydrostatic forces below ground which tilt the EC scales when used in low-rise construction.

The below diagram indicates some typical project scores of new-build constructions. As previously established, there is a wide range of upper and lower bound outcomes related back to design and specification decisions over the course of these projects (Fig. 9).

The below diagram indicates some typical project scores of refurbished and adapted buildings. From a structural perspective these reinforce the belief that it is more sustainable to keep the shell-and-core of a building and adapt and re-use

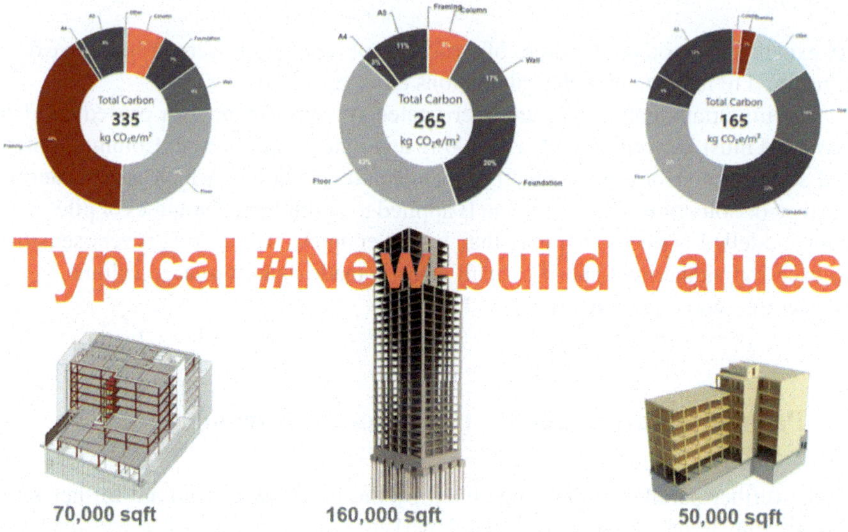

Fig. 9 Typical new build example scores

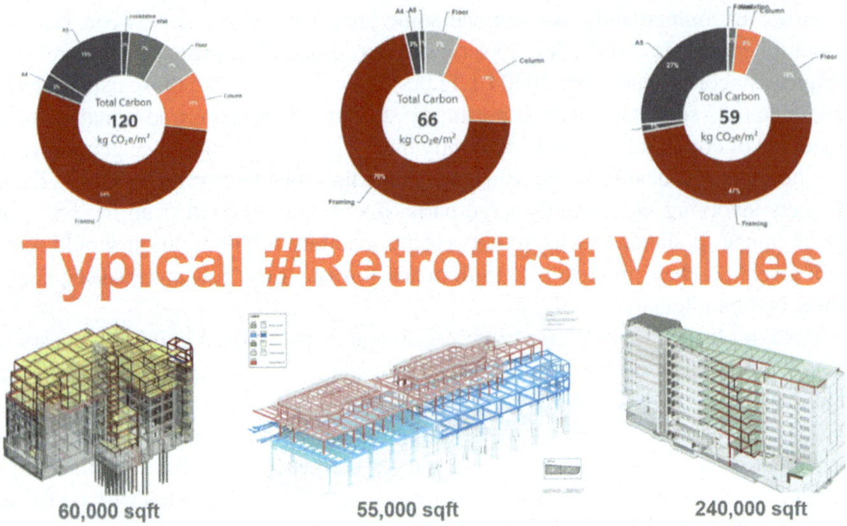

Fig. 10 Typical Adaptive Re-use example scores

rather than demolish, as across the board they are only a fraction of the new-build values (Fig. 10).

6 QA Process

To expand upon potential sources of error a couple of example screenshots have been illustrated to indicate how the calculations are scrutinized.

It is important that outputs are interrogated—much like any QA procedure, data retrieved must be verified with a checking and quality assessment. Common potential sources of error are incorrectly modelled Revit families which need materials corrected. Sometimes reinforcement is applied to unreinforced volumes or piles have been modelled as steelwork rather than In-situ concrete. The scoring is very sensitive to each individual item and if a small error is made against a large volume, it can impact the score dramatically (Figs. 11, 12).

7 Parallel Studies Relating to Embodied Carbon

Some further parallel studies completed by the EOC R&D team are further reinforcing the trends seen in our benchmarked project scores when we compare Like-for-like structural materiality against EC.

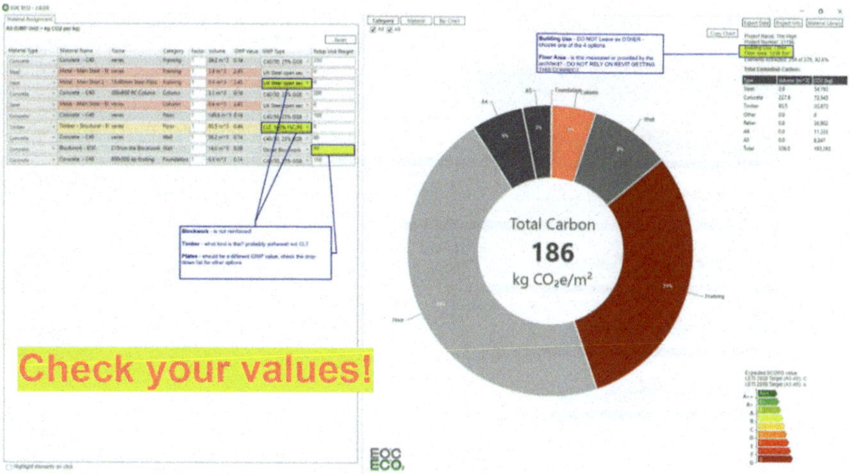

Fig. 11 Example no.1 QA comments from case study project

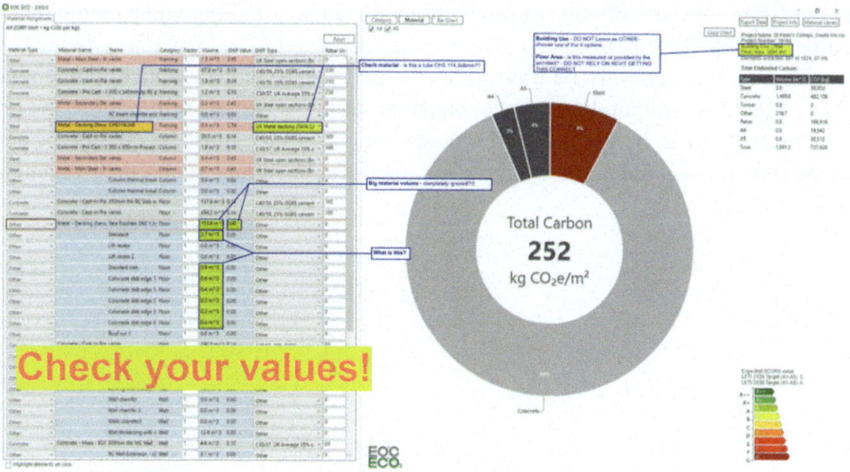

Fig. 12 Example no.2 QA comments from case study project

As illustrated in Fig. 13. any building with a structural column grid above 10 × 10 m on plan may struggle to achieve LETI 2030 targets for structures as a whole so there is certainly an incentive to help clients and architects understand these simple metrics.

The Future of Concrete?

Further parallel research completed by EOC indicates that of the traditional concrete options, recently considered the 'bogeyman' of the construction sector due to the

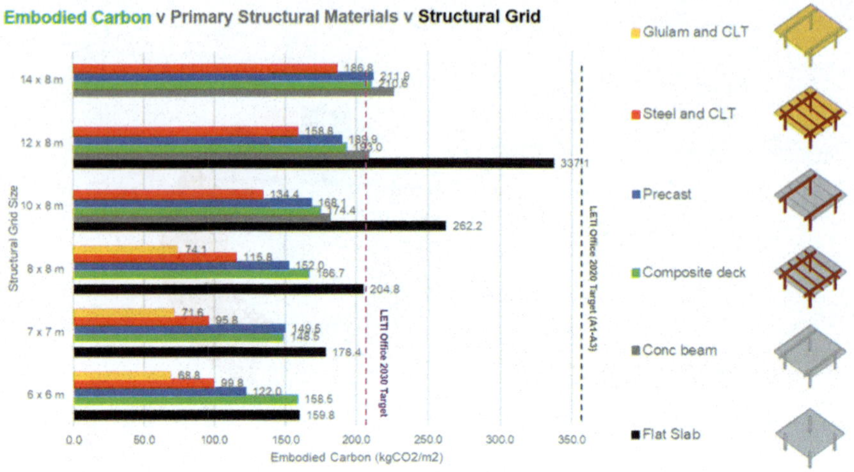

Fig. 13 A1-A3 Stage Embodied Carbon by Primary Structural Material versus Grid Size

high carbon intensity of cement production, are not so polluting when compared to current targets (Fig. 14).

Concrete can certainly be used to hit the LETI 2030 targets whilst maintaining fairly generous structural grid sizes, but some of those options will require deeper structural zones to be granted and therefore this message must be communicated to the wider stakeholder, the design team, the local government and their planning

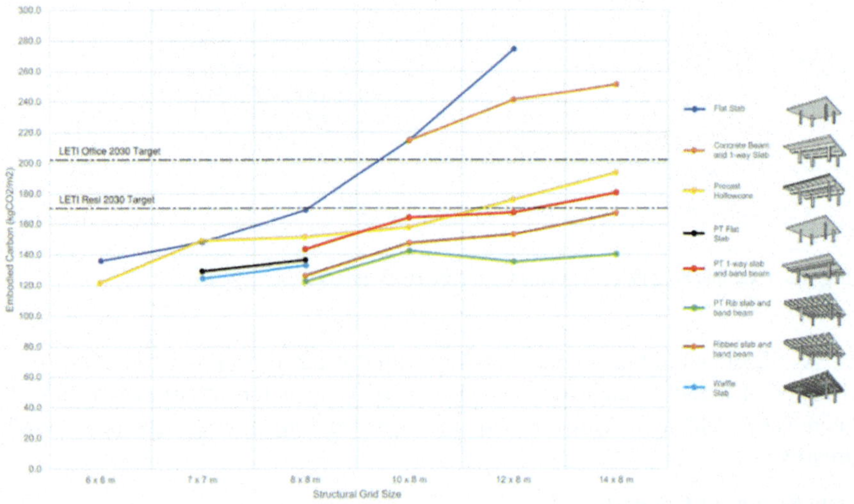

Fig. 14 Stage A1-A3 embodied carbon for concrete frame typologies versus grid size

departments, to take away the pressure related to adopted massing, overall building and storey heights.

It is also noted that the emerging specification of cement-free concrete will further improve the strategy towards future use of concrete however we will still need to measure and record against the available targets, so improvements made across the supply chain need to be fully assessed and understood.

8 Conclusions and Next Steps

The carbon intensity values we use for our projects will be updated annually on a rolling basis to maintain relevance to the continually changing power grid. The monitoring of these factors is a life project and significant time needs to be invested to get full disclosure on our industry output. The development of EOC ECO_2 will assist with the rapid extraction of data which we need to assess construction on a national and international scale.

Whilst LETI and RIBA have defined embodied CO_2 targets for buildings in the UK there is less discourse on infrastructure, bridges, roads and tunnels and therefore a significant quotient of the construction sector does not yet have a defined limit. If limits are yet to be set the EOC ECO_2 tool could be useful in collating the data for those wider industries.

The more data we can retrieve the wider the sample size we have and the more variables we can record for each case study the better we are able to assess.

EOC ECO_2 has allowed Eckersley O'Callaghan to calculate the embodied carbon of past and current projects and benchmark them against targets such as LETI & SCORS, helping them develop sustainability strategies for projects while having the data and knowledge to accurately advise clients on structural approaches that need to be adopted to work towards achieving Net Zero Targets. As a result, we have been able to discourage unnecessary demolition and quantify the advantage of maintaining existing structures.

The tool has in-built flexibility to allow it to be adapted for any supply chain or any project location anywhere in the world. The user can simply define their own bespoke materials library or choose one of the readily available 'stock' materials already uploaded within the software.

Using EOC ECO_2, Eckersley O'Callaghan have already developed a valuable data set that helps us make informed decisions on projects on how to specify and use materials, and ultimately achieve more sustainable outcomes. The ability to communicate more clearly about sustainability, a topic that is often left to marketing, allows them to make decisions while being conscious of the large impact the industry has on the natural environment.

Eckersley O'Callaghan believe that nobody should be allowed to own the 'IP' on sustainability, and that's why they decided to share the software open platform. The ripple effect has been widespread. Globally over 150 organizations from structural and facade engineering, architectural companies to product designers have begun to use the software. There are no freely available products on the market that compare with this tool. Climate change is a global problem and only by sharing patent free software will a better understanding be reached.

9 Biographical Notes

Duncan Walters, 2003 graduated with a BSc in Automotive Design, moved into manufacturing spending several years as a machinist and precision engineer developing valuable practical skills which benefitted a move into building design, completing a Civil Engineering MSc at UCL in 2007. He joined Eckersley O'Callaghan in late 2014 and is now a Director in the Structures Group.

Chiara Bariviera is a Senior Engineer in the Structures Group. Having achieved First-Class Honours in her Bachelor of Civil Engineering also at UCL, joined Eckersley O'Callaghan in 2018. She balances her time between the production of structural design material for many prominent projects and leading the aspirations for improvements in monitoring and reducing carbon budgets within our live projects, tracking the data from past projects and consulting on best practice for future developments. Chiara also developed an early concept stage Carbon and Energy Tool to assess and quantify embodied carbon and energy emissions in building elements. Additionally, she designed a timber spreadsheet for quick assessments of capacity in screwed connections using data from CLT and BauBuche manufacturer's guides. Her work now spans all sectors and scales of building design projects.

Reference

1. CLF Homepage, https://carbonleadershipforum.org/embodied-carbon-101, Last Accessed 22/06/2023

Chapter 3
Interoperability Challenges. Exploring Trends, Patterns, Practices and Possible Futures for Enhanced Collaboration and Efficiency in the AEC Industry

Alessio Lombardi

Abstract The AEC industry faces significant challenges in software development, including reusability, scalability, and interoperability issues stemming from its diverse and segregated nature. Multiple domains and subdomains within the industry create coordination difficulties when striving for a unified design. Existing software solutions inherit these challenges, often relying on individual products and plug-ins built on proprietary software. These solutions lack interface and interoperability, necessitating users to comprehend different underlying assumptions. Furthermore, limited accessibility to code bases hampers collaboration for non-authors due to suboptimal software engineering practices. Open-source solutions like BHoM and Speckle offer promising approaches to enhance interoperability by enabling data interchange and concept mapping between domain-specific vocabularies. However, their widespread adoption and scalability rely on industry-wide acceptance. This chapter provides an overview of software development approaches in the AEC industry, focusing on recent open-source interoperability solutions. It examines knowledge representation, design patterns, trends, and practices, while discussing potential strategies for addressing software engineering challenges in the AEC industry.

Keywords Parametric modelling · Interoperability management · Domain-specific challenges · Fragmentation · Software solutions

---

A. Lombardi (✉)
Buro Happold Engineering, 230 Lower Bristol Rd, Bath BA2 3DQ, UK
e-mail: alessio.lombardi@burohappold.com

P. Ruttico (ed.), *Coding Architecture*, Digital Innovations in Architecture, Engineering and Construction, https://doi.org/10.1007/978-3-031-47913-7_3

1 Introduction

1.1 AEC: A Fragmented Industry with Many Software Solutions

Software engineering and, more generally, information technology for the Architecture, Engineering and Construction (AEC) industry reflects the complexity of the industry. As with any industrial sector, the aims of software development are to solve the domain-specific challenges presented by an industry. However, the AEC industry itself is multifaceted, with domain experts coming from extremely different backgrounds often working together on the same product. Each discipline uses a specific terminology, and even the target outcomes and specific responsibilities of each discipline are sometimes complex to define. A global economy adds to the complexity of the sector, as an increasing number of projects are undertaken by companies from different parts of the globe. These facts reflect on the software products and software development disciplines put into practice in the AEC sector and translate into their own technical challenges.

The variety of the AEC industry differentiates it from other industries also from the perspective of individual parts composing the final product. In cases like the automotive industry, the assembly has generally a good delineation of parts and final product that is generally also well reflected on the producers and designers' responsibility. The intrinsic complexity is therefore also reflected on the software used and applied to solve the many challenges in the AEC industry, which have to deal with reusability, scalability and interoperability issues.

1.2 Fragmentation, Domains and Terminology

In AEC, concepts have several ways of being identified, depending on the domain-specific perspective. For practical reasons, the AEC industry is in fact often subdivided into a multitude of segregated domains and subdomains—architecture, interior design, structural, acoustic, environmental, mechanical, electrical and plumbing engineering, fabrication, construction, and many others. Several challenges arise when these areas need to coordinate and interoperate to achieve a single, coordinated design, and multidisciplinary solutions are always required [1]. This frequently leads to fragmentation [2], where a single physical asset is represented in many distinct disciplinary representations that often lack a robust and easily retrievable connection between each other [3].

1.3 Producing, Storing and Exchanging Information

The fragmentation of the industry has repercussions on the channels and storage systems for information and the data that represents it. The design of a building comprises models that represent and analyze every specialized aspect of it, and such models are produced with a variety of different software, each dedicated to a specific discipline. Whilst the architectural representation of a building often begins from Computer Aided Design (CAD) software as simple two-dimensional drawings, the design eventually will evolve into a more complex Building Information Modelling (BIM) representation that is capable of including not only the architectural representation of the space, but also of its functionality. Mechanical, Electrical and Plumbing (MEP) elements will be incorporated; the sizes of the structural elements such as columns and beams will be refined via the structural engineering design; the windows and walls will reflect the energy analysis of the building, and so on. However, all the analytical models produced in the effort of reaching this holistic representation will end up being unlinked with the final representation itself. Further, the fragmentation of the information into separate models requires costly interoperability management.

1.4 Software Solutions to Design Challenges

Current software solutions often adapt to pre-existing hierarchies, often inheriting the above-mentioned issues of the AEC industry. In AEC exist many stand-alone, highly specialised software products that are designed to deal with domain specific problems. However, these domain-specific problems are often interrelated and can't be fully resolved in isolation. This means that iterations are required in order to ensure that the independently developed solutions are compatible with each other. Observing the features, the tendencies and development practices behind the different software solutions can be helpful to suggest what the future could look like and how the industry may grow with them.

1.5 Parametric Modelling and Visual Scripting: Focus on the Right Problem

Parametric design is often associated with Grasshopper and Dynamo, two plug-ins for licensed software dedicated to geometrical modelling and BIM. This is primarily because the term design is often associated with the aspect of the building model that can be seen, i.e. geometry, and those two tools were initially developed to solve problems in their area of origin (geometry and BIM). However, parametric design can also refer to other specialised tasks, e.g. structural engineering optimisation, solar light simulation. The fact that Grasshopper and Dynamo implemented the visual programming paradigm—i.e. components representing functions, and input and outputs transferred between components via visible "wires"—contributed to extending their success outside of the geometrical and BIM domain. Grasshopper is in particular largely responsible for the success of visual scripting in the AEC industry, to the extent that visual scripting is now largely associated with parametric modelling, and vice-versa.

An association that is less frequently made is the similarity between other software products and visual scripting tools like Grasshopper, in particular spreadsheet software like Excel. Excel is ubiquitously used across the industry—properly and improperly—to deal with a wide variety of problems, from simple calculator to database system. The similarity becomes striking when the relationship between cells is visualized. The cells containing formulas are equivalent to components or nodes in Grasshopper and Dynamo, whilst cells containing data are like parameter nodes. However, the difference in user experience is notable, and may explain in large part the difference in applications of, e.g., Excel and Grasshopper. In particular, Grasshopper has become the building foundation for many other extensions and plug-ins that allow it to deal with domain-specific problems of the AEC industry.

1.6 Extensions and Plug-Ins to Solve Domain-Specific Problems

In order to facilitate the widest possible variety of workflows, software developers and AEC practices often develop plug-ins and extensions built on top of proprietary BIM and parametric design software. For example, FoodForRhino is a large dataset that collects plug-ins for Grasshopper, and a wide variety of plug-ins exist for many different domains, from FEA modelling (e.g. Karamba3D) to Sunlight and Energy simulations (e.g. Honeybee, LadyBug tools) and many others. These extensions in many cases grow to become a key part of parametric workflows implemented by domain-expert designers. Spreadsheet software like Excel and Google Sheets are also targeted by several extensions, but their scope is often narrower and more oriented towards data management, rather than to deal with domain-specific problems, a fact that can be explained with the more focused user experience that spreadsheets can expose.

Extensions and plug-ins are regarded as a convenient solution to solve design problems, especially at the early stages of design. However, they pose a set of problems. First, extensions built for the same software rarely interface or acknowledge each other, leaving interoperability to the user or to the proprietary software that they extend or leverage. This reinforces the dependence on the parent software, a fact that in turn can generate scalability issues, versioning issues, and vendor lock-ins. Second, each extension comes with a set of conventions, assumptions and limitations, and requires designers to be aware of them and learn how to use them. Finally, often plug-ins are the product of individual efforts, like single software developers, or small distributed teams as side activities. In these cases, plug-ins are commonly released to the public free of charge, either in Open or Closed source form (i.e. sharing or not sharing the source code for other people to contribute to it), or on a freemium basis.

Plug-ins and extensions also may suffer from problems deriving from the way they were developed, which add a short- and long-term cost on users. Because they are built to answer specific problems and without the planning of a business software release, and because they are frequently developed by people from diverse backgrounds, extensions thus produced are not always built following software engineering best practices. This means that the resulting code bases may be hard to work with for non-authors or late comers, scale up poorly, and are generally difficult to maintain. At the same time, because plug-ins generally target users that are non familiar or likely not interested in contributing to their development, there is little incentive to create a widely understandable codebase that may be easier to share and contribute to by a wider audience. Therefore, unless the original authors remain active on their work, there is always a high risk that these extensions become abandoned, unsupported or out-of-date. This then reflects on the work of designers, whose scripts based on such extensions may become deprecated or out-of-date, and therefore the cost of software maintenance and usage becomes shifted to the end-users themselves. Food4Rhino shows many examples of once successful plug-ins for Grasshopper whose last release is several years old, some of which were made open source (OS) by their authors—that is, their codebase was made public and freely modifiable—in an attempt to spark community interest in their maintenance, with little to no success [4]. However, other tools which shared their codebase as part of a structured OS strategy have had wide success and thrive.

1.7 Information and Knowledge Modelling

Even an agreed representation of information, concepts and how they can be modelled into data is challenging in the AEC industry. How information is encoded into data defines the possibility of doing interoperability; interoperability can be described as the possibility of communication, data exchange and use between different software. In industry and academia, different frameworks and data schemas are developed and being used to tackle the interoperability problem.

1.8 Industry Foundation Classes

Industry Foundation Classes (IFC) is a standardized data schema for the AEC industry which can be used to store information throughout the lifecycle of a built structure [5]. IFC was developed along with a set of building product models by the International Alliance for Interoperability (IAI) (former name of buildingSMART) to represent and exchange data in the AEC industry. IFC mainly targets the referencing or archiving of model information, with little interest in how the model is created, in information querying, investigation or operational processes. Rather, IFC aims for a comprehensive, holistic representativeness of the building information, and it is powerful in representing geometric data, element classification and product properties. However, IFC falls short when it comes to representing dynamic data—data that is frequently updated, as it happens during the early stages of design, due to the fast and frequently changing design needs.

IFC data schema consists of four layers: domain layer, interoperability layer, the core layer, and resource layer [6], where each individual schema is assigned to one conceptual layer. The schema represents building elements including semantic information such as attributes or relationships, as well as abstract concepts or processes. The latest IFC schema version, IFC4 $\times$ 2 has 407 types and 816 entities. While the intention of modularizing the schema was originally there, the IFC schema resulted in multiple dependencies between concepts and modules, bringing it in the opposite direction [7].

Industry Foundation Classes (IFC) provide a rich schema for interoperability that can however be redundant and ambiguous. IFC lacks semantic clarity in mapping entities and relationships, resulting in the potential for different data structures to hold the same information. Because of the multidisciplinarity of building design, several representations for the same concepts, seen from different lenses, are often necessary. However, collecting these representations under a single large schema poses challenges and brings inconsistency and ambiguity risks. From a Software Engineering perspective, IFC is strongly object-oriented, and uses inheritance as a fundamental strategy, a fact that hides inconsistencies when observing the specification alone. It's easy to find different classes that may well be used to represent the same concept and hold similar data; in this sense, the distinction between some IFC classes is not well defined [8].

1.9 Semantic Web Standards, Linked Data and Ontologies

The Linked Building Data (LBD) academic community group has been researching on extending semantic web standards to the AEC industry. The Semantic Web Standard was born as an extension of the Web originally proposed by the World Wide Web Consortium (W3C), in order to promote common data formats and exchange protocols on the Web, with the final aim of allowing data "to be shared and reused across application, enterprise, and community boundaries", data that is easily processable by humans and machines [9]—Linked Data. Linked Data follows standardized ways not only in defining schemas and terminology, but also in the way the data is published and linked. The Linked Building Data-Community Group (LBD-CG) is using Semantic Web standards as an open and decentralized alternative to the existing centralized and file-based approaches to storing and sharing data.

Linked Data leverages the so-called Resource Description Framework (RDF) which, in essence, outlines how every entity, or concept, can be linked to one another via triples (subject, predicate, object). A set of triples, which interconnect several concepts, can be defined as Ontology. There exist different languages and means to define such triples, and most of them leverage a fundamental work called Web Ontology Language (OWL). OWL has a rich expressivity and is supported by several reasoning tools, which allow to perform inference and information querying from existing data [10]. OWL allows for a rich description of the data in terms of concepts and relationships between concept individuals which build on the RDF framework.

Many different AEC ontologies have been proposed. For example, IfcOWL ontology expresses entities, types and properties based on the IFC Schema [11]. While most LBD ontologies are meant to be modular and focus on a specific domain, so that they are easy to connect to other ontologies, IfcOWL is instead an extensive schema that overarches several domains, making it harder to relate to other ontologies. Another famous example is Building Topology Ontology (BOT), a minimal ontology for describing the core topological concepts of a building [12]. A good example of a highly specialized ontology in AEC is the Teddy Ontology (TDY), created as part of the SOFiSTik structural engineering package, which addresses structural analysis related attributes of building elements.

There are other approaches that build focus on the interconnectivity of concepts to facilitate interoperability. One example is Brick, it is an open-source curated and extensible dictionary of terms (a schema) and relationships for linking concepts. Brick easily integrates with BOT, as both leverage the Semantic Web. Another different approach, which instead uses a tagging mechanism to add information and link data, is Project Haystack, which aims to describe building assets using semi-structured sets of tags. However, Haystacks doesn't provide a framework to build and use tags, so it is more geared toward quick data management and querying rather than consistent modelling of data.

2 Community Sourced Co-Design and Interoperability: BHoM and Speckle

Co-design considers design and analysis methods, manufacturing and construction processes, as well as material and building systems simultaneously; it does so by suggesting that integration of data coming from different disciplines should start from the early phases of design, which are the ones that weights the most on short- and long-term costs. However, in practice, this integration is often hindered by interoperability problems between different software. Combining multiple knowledge representations is still a significant challenge that has not been entirely solved so far [13], and recently some grassroots open-source frameworks have been targeting this problem.

2.1 Speckle: Data Interoperability and Sharing

Speckle is a data communication web-based platform which is built focusing on 3D data. As such, it allows real-time collaboration, data management, versioning and automation of geometrical data. It was initially developed as an academic project by D. Stefanescu at UCL, as part of the InnoChain project, an H2020 Marie Curie European Training Network. The mission of Speckle is described as "to enable everyone to speak in 3D", therefore the primary attention is given to geometrical data which can be enriched with semantic meaning, rather than the other way around. Speckle offers a library with about two hundred object models, a web viewer for geometry, and dozens of connectors towards different software. However, its main focus is on connecting software and exchanging (primarily geometrical) data, rather than representing building elements.

Speckle leverages a server-based architecture, where a Speckle server deals with data transfer between different endpoints. Speckle functionality is grouped in Speckle Connectors and Speckle Kits. Connectors are plug-ins for specific software that allow the software to receive and send data through Speckle servers. Instead, a Speckle Kit is a collection of a specific Object Model (a schema), and of Conversion functionality that allows to interchange data with other software. Each Kit is assumed to be able to handle all Speckle types [14]. Additionally, when a connector needs to convert an object, it will explicitly pick what Kit to use, if more than one is available. Whenever a connector is "sending to Speckle", it needs to convert data (lines, points, beams, doors, etc.) into a Kit's intermediate object model. Upon receiving, the opposite conversion happens (Fig. 1).

This approach is software-centric, and comes with pros, but also several cons. This organisation of Kits is lean and modular and allows to start supporting functionality for a specific software rapidly. However, Speckle Kits will individually have to make sure that the conversions work for every software of interest: for example, if both Kit X and Kit Y need to target Revit, then Revit conversions will need to be written for both Object Model X and Object Model Y. Users writing a Kit that targets a specific domain or software may not be familiar with Revit, and this requires each Kit to have a wide scope.

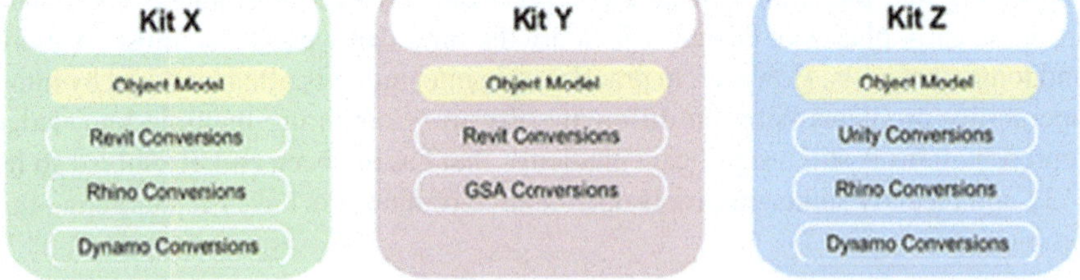

Fig. 1 Structure of speckle kits

2.2 Speckle Schemas

In order to reduce the number of ad-hoc conversions and method calls, a central object schema, called Objects Kit, was created. The central schema can contain conversions and objects that are common to several kits. The classes in the Objects Kits are described as Data Transfer Objects (DTO), a software engineering term that describes classes whose main responsibility is to carry data across processes, in order to reduce the number of method calls, and that are serializable [15]. The main schema is divided in two categories: "Geometry" (points, lines, and so on) and "Built Elements" (anything tangible in the Built environment, such as ducts, beams, openings, and so on). If a Kit intends to extend the schema, it can do so by defining classes within a specific category, for example Objects.BuiltElements.Revit will contain Revit Built Elements. This requires a distinction between Geometry and Built Element to be made at class definition. Any speckle class should inherit from the Speckle Base object, that hosts some default functionality. For example, a Base object can be dynamically extended with dynamic properties. Dynamic properties are set via C# dot notation and can be retrieved similarly to Dictionary indexing [16]. This allows to mix Static typing (often improperly referred to as "Strong" typing [17]) and Dynamic typing on Speckle objects.

When defining classes, specific host application support must also be embedded into the schema itself. For example, if a class is to be discoverable in Grasshopper, then must be signaled via C# attributes within the class definition itself, i.e., in the constructor [18]. This mixes the schema's data representation, or semantic power, with software support information, and further complicates the schema definition. Because schema definition is likely to be made by specific domain experts, this mixed responsibility of classes may well pose a significant obstacle to scalability and ease of definition (Fig. 2).

Versioning of the schema is also handled within the schema itself. If changes are done to a class, a dedicated Attribute (SchemaDeprecated) needs to be added in the constructor. This adds another layer of non-semantic responsibility in the class definition itself.

Also, the management of the data is handled in the schema. Speckle calls this Decomposing and Composing. By adding a specific attribute to a class property

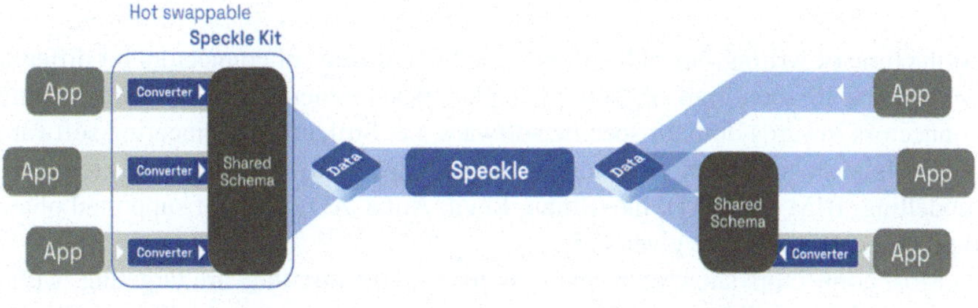

Fig. 2 Each speckle kit will define converters for each app it wants to support

(DetachProperty), that property's data is marked to be stored as a reference. This adds further responsibility on the class definition, although it removes the burden of defining a mechanism that takes care of storing a reference rather than the data for specific properties: this is deferred to the Speckle server instead. The recomposition of an object happens within the deserialization process, as part of the transport layer. Speckle receives the json of the object, retrieves the reference tree of the object, and de-serializes and re-composes the object, inserting actual referenced object in place of references.

2.3 Managing and Querying Data

Speckle organizes data in Streams. A stream is simply a data collection, described with some additional information useful to manage it, and permissions for read and write. Each stream is uniquely identified with a StreamId, which is assigned by the Speckle Server. The data in streams can be further organized into branches which can be modified via commits, an approached borrowed from Git [19]. Streams are the main mechanism by which data is shared between people and applications; a stream could be created in Revit, from where data could be sent, and then the data could be accessed via the Speckle Web App, or received from a connector in Grasshopper. Streams expose also events, which can be monitored via WebHooks, e.g. for automating functionality whenever a stream is modified [20].

In general, Speckle assumes that a Speckle server is present wherever it runs. A speckle server can be spun off on the local machine, on a server in an internal company network, or in the cloud. Authentication and access to the data can be managed via API tokens [21]. Additionally, an abstraction layer for data management is provided, called Transport. This is a generic interface to a backend storage solution, like SQLlite, disk, memory or others, which allows to decouple from the Speckle Server; this functionality is currently WIP [22]. Speckle server data can be queried via a REST API and via a GraphQL API, which allows for complex querying [23].

2.4 Speckle Connectors

At the time of writing, Speckle offers a couple of dozens of connectors for different software, with about half of them in "alpha" development stage [24]. Most of the connectors towards domain-specific software, i.e. Structural Engineering software, are in alpha stage. The most mature connectors appear to be those in the area of 3D modelling, BIM and CAD, like Rhino, Revit, AutoCAD. Table of supported object types per connectors are given [25].

Most connectors show up as plug-ins for the software they are targeting, with a dedicated UI that allows to send and receive objects from a Speckle Server instance [26].

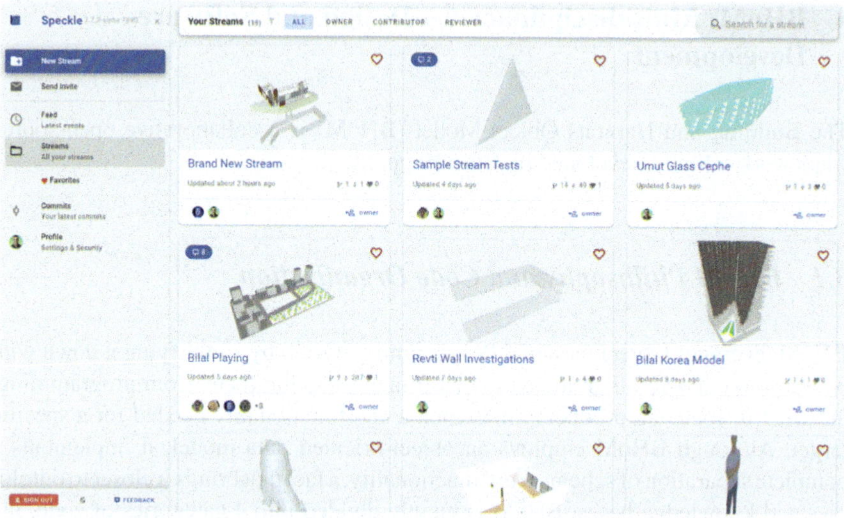

Fig. 3 Speckle web app interface

2.5 Speckle Web App

A strong selling point of Speckle is the Web App with embedded 3D viewer. This allows to manage and coordinate data, viewing/editing permissions, and view the 3D geometry associated with the data. A Web App instance can connect to a specific Speckle Server and manage streams and users present on the server. The powerful 3D viewer allows for many different geometries to be visualized (Fig. 3).

2.6 Code Management

Speckle arranges the code into base repositories, whilst kits are in dedicated repositories. At the time of writing, the main base repository, Speckle-sharp, contains a mix of schema [27], conversion functions [28] (https://github.com/specklesystems/speckle-sharp/tree/main/Objects/Converters), and connectors [29]. This mixed codebase makes the repository the target for issues that pertain completely different worlds, for example, Structural software and ArchiCAD [30], and may be deemed inefficient in crowding the best expertise in the right place. Kits stored in separate repositories, like Unreal [31] seem to be placed separately for modularity (a kit should be "hot-swappable" [32]), whilst the logic to place software connectors within the main Speckle-sharp repo seems to be related to what is deemed to be an "authoring application" [33]—although this definition may seem arbitrary.

3 BHoM: Multidisciplinary Co-Design and Software Development

The Building and Habitats Object Model (BHoM) is a collaborative open-source framework initially developed by Buro Happold.

3.1 BHoM Philosophy and Code Organisation

BHoM leverages Object Oriented Programming (OOP) by simplifying it down with patterns and an overall architecture that brings away the focus from programming, allowing to focus on producing the schema and functionality needed for a specific target. Although BHoM employs an object-oriented data model, it implements a complete separation of schema from functionality, a fact that brings it closer to ontologies and knowledge bases. BHoM is mainly divided into 4 categories of code: the schemas (Object Model, oM), the functionality (the Engines), the connectivity (the Adapters) and the User Interfaces (authoring software, UIs). Whilst the core functionality and main schemas are hosted within base repositories (BHoM, BHoM_Engine, the rest of the codebase is organized into separate, modular and "hot swappable" Toolkits (Fig. 4).

Whilst this approach may have an initially steep learning curve, the complexity on the long run is reduced and retains manageability, if compared with other frameworks. This is also due to the many limitations imposed by design on the framework itself, which enforce simplicity and consistency. On the other hand, the system may be prone to rigidity; however, any large schema model does require some degree of stability to prove truly useful. Additionally, the Open Source, collaborative approach

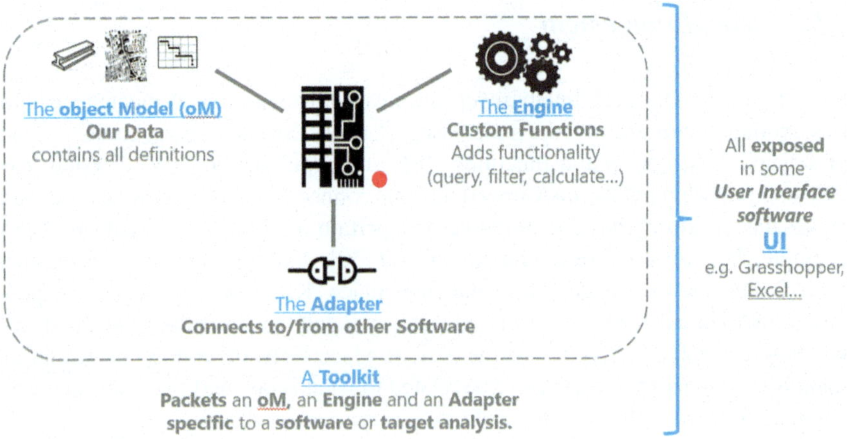

Fig. 4 BHoM code organization

that BHoM is founded upon encourages experimentation and the lean creation of extensions—Toolkits—in which functionality can be rapidly created and tested.

3.2 BHoM Schema: The Object Model

BHoM schema is made by C# classes. Classes are grouped per disciplinary area, and each group is called Object Models (oM); each Object Model is identified by a dedicated namespace. For example, there is an Object Model for Geometrical Types, called Geometry_oM, that exposes class within a BH.oM.Geometry [34] namespace; Structural Engineering owns Structural_oM and the BH.oM.Structure [35] namespace; architectural concepts can be found in Architecture_oM within the BH.oM.Architecture [36] namespace, and so on. Each object model is stored in a dedicated C# project, which is in turn collected in a specific Visual Studio solution. For example, the core schema that collects all the main classes is called BHoM and contains a Visual Studio solution with 26 projects (and related namespaces) and hosts about 850 types at the time of writing.

Object models only store classes which have no functionality (i.e., no methods) and are composed only by public properties. In this sense, BHoM strays from classical OOP, which defines objects as instances of classes that owns both data and functionality; a more apt terminology for BHoM classes would be DTO (Data Transfer Objects) or Object Relational Mapper classes (ORM) [37]. Regardless, this kind of organization allows to separate concerns of developing the right representation for a concept and the functionality that the concept should be subject to or expose.

Additionally, BHoM exposes certain rules for how the concept should be defined in a schema. In general, a class should define only orthogonal properties, that is, properties that do not repeat information between each other. For example, a Column class may define a Start Point and an End Point as separate properties, but then it cannot define a third property called "Line" which goes between a start point and an end point, as it would be redundant. Non-orthogonal properties may be defined by making it explicit that the object may contain redundant information [38]. Furthermore, properties on the class should always expose public get and set accessors, and no private fields are allowed in a class. This is to ensure that the schema is clean and does not contain any kind of functionality.

This approach in defining the schema brings several benefits and some disadvantages. First, when defining a class, the domain experts can crowd around the area of expertise they are most familiar with, e.g. Structural Engineering, only requiring exposure to the concepts they need. This means that defining new classes for certain concepts can follow a strategy that is more similar to devising a dictionary entry definition than to actual software development, which in turns expands the user base capable of creating new types. On the other hand, users should be familiar with the large schema that BHoM exposes to know which classes may or may not make sense to create; however, BHoM encourages experimentation in separated Toolkits, where users are free to create classes as they see fit. When a new class is well settled in a

Toolkit, stable and established within a specific workflow, then an investigation can be started to considering porting it to the core Object Model schema, if needed.

Secondly, class definition owns no responsibility other than describing a good schema for the concept that the domain experts want to represent. Unlike Speckle, data storage, versioning, and serializations are all decoupled from the schema definition. For example, a separated versioning mechanism ensures that if a property changes name, or if a class is renamed, the compatibility is kept between different BHoM releases, without requiring attributes to be placed on the properties or class themselves [39].

Further, because each class owns only data, all classes can be exposed directly in authoring software and user interfaces. For example, a Column object can be created in Grasshopper, Excel or other user interfaces supported by BHoM, as well as from other C# runtimes and programming environments.

3.3 BHoM Functionality: The Engines

Similarly to oM projects, Engine projects target a specific domain grouped within specific namespaces e.g. methods for Structural Engineering are placed in the Structure_Engine under namespaces starting with BH.Engine.Structure. Engines are essentially collections of five different kinds of static classes, which are used as groups for the functions (Table 1).

Each method must clearly target one main input type. BHoM's runtime, which is assumed to be loaded in any User Interface software and environment that is used to author BHoM objects, exposes a dynamic binding that associates methods with the object type that they target, so that methods are available throughout the framework as an extension to the type. Additionally, methods are defined as Extension methods [40], which allows to show the functionality as an extension to the objects even when programming. This makes the oM C# classes work in a manner that is closer to dynamically typed languages (e.g. Python) than statically typed languages, and is a design choice that simplifies contribution and scalability. Each oM type can effectively be augmented with precise functions included in Engines.

Table 1 Engine methods

Engine method static class	Description
Create	To provide ways to instantiate objects (classes should not define constructors, unless they are Immutable)
Modify	To modify objects
Query	To retrieve derived data from objects
Convert	To perform conversions between different object types
Compute	To perform operations that may be computationally demanding, with much I/O, or that don't fit any of the other categories

An interesting consequence of this design pattern is that Query methods in BHoM_Engine can be used to define derived properties of classes. For example, the centreline of a Bar [41] (a linear element used for Structural Analysis, which can be used to describe beams and columns from a structural engineering perspective) is an example of a derived property. The Bar is defined with a Start Node and an End Node; the line that connects the two nodes is not defined as part of the class, so to maintain the orthogonality of the class. The centerline can however be derived from a Query function; this is a function that returns the centerline of the Bar as the line between the Start Node and End Node [42]. The Centerline function effectively extends the capability of the bar, without polluting the schema, which remains dedicated to storing the semantic representation of the concept of a Bar.

3.4 Connectivity: The Adapters

Adapters host the conversion functionality between the object models and different software. Each adapter is responsible for invoking the conversion functionality needed to map a BHoM schema to the software equivalent, if present, and to invoke any required API functions to export the data to the target software. The schemas are therefore free to represent the concepts without concerns about the final end software representations, as this responsibility is shifted to the adapters. This creates an ecosystem where, once an adapter for a certain software is created, the central object model is free to be exported and imported to and from all other software supported by BHoM. Compare this with, for example, Speckle, where conversion methods between the same software have to be specified across different Kits. Currently, BHoM hosts over 30 adapters, in varying development state. The most mature adapters are those in the Structural Engineering disciplines (e.g. GSA, Robot, ETABS), Environmental and BIM (Revit) (Fig. 5).

3.5 User Interfaces

BHoM's class and functionality organization make it directly compatible with both flow-based and text-based programming. User interfaces expose the schemas (oM), functionality (Engine) and connectivity (Adapter) into three separate categories, from which users can access all of the codebase. For example, any method can be accessed in the same way from any User Interface. In this sense, BHoM demonstrates the similarity between spreadsheets and visual programming environments such as Grasshopper; for example, in Excel, a BHoM function correspond to a cell containing a formula, whilst an object can be stored in a data cell. The same function appears as a Component in Grasshopper, as a Node in Dynamo, and so on (Fig. 6).

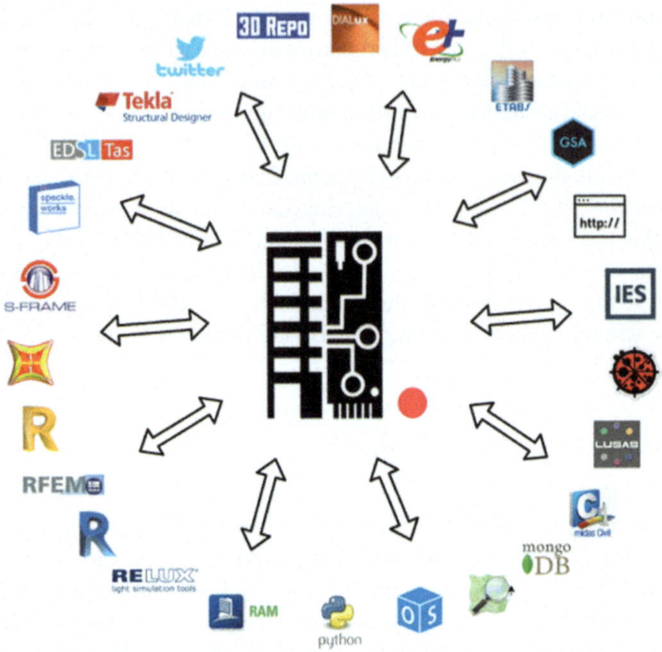

Fig. 5 The Adapters connect the central schema with different software with a 1:1 connection. Rather than software connected to software, BHoM connects information with different representations of itself

4 What's in the Future of AEC?

A grassroot, open source, collaborative approach to data sharing, versioning and software interoperability.

During 2019, the "AEC Delta Mobility" project, funded by Innovate UK under the "Increase productivity, performance and quality in UK construction" grant, explored how to streamline data workflows in manufacturing. In particular, the target was to engage manufacturing in the earliest possible stages of design, by streamlining data workflows from consultants to factory [43]. Key contributors in the success of the project were the proposed integrations of BHoM, Speckle and 3DRepo, another software solution to manage and share BIM data. The developed prototype Adapters for Speckle and 3DRepo demonstrated the functionality and advantages that a coordinated development could bring to the industry. Combining the strong data transferring and management functionality of Speckle and 3DRepo with the strength of the consistent and multidisciplinary schema of BHoM could prove as a winning strategy for a real grassroot interoperability software solution in the AEC industry. Whilst the pandemic and reorganizations hindered faster integration of these tools, it would be desirable for the industry and the Open Source community to crowd around and merge the strengths of these valuable software solutions. The resulting collaboration

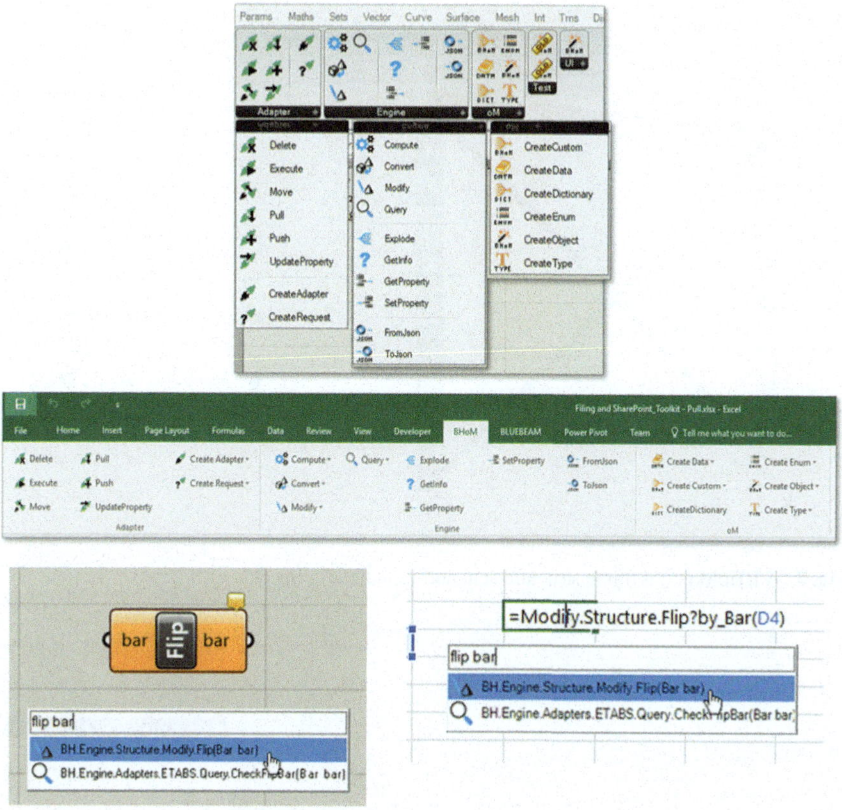

Fig. 6 BHoM functionality exposed in different user interfaces

could well prove strong enough to shake the industry and establish the open-source paradigm and grassroot nature of software solution as a successful factor to truly create innovation, improve efficiency, and open up new ways of working (Figs. 7 and 8).

4.1 Knowledge Management and Inference Reasoning in AEC

Multidisciplinary co-design requires multiple domain experts to crowd around and work towards the same outcome; different terms that refer to the same thing seen from different perspectives exists. However, connecting the data represented in different ways would bring several advantages, like cross-disciplinary information querying, validation and clash detection. Furthermore, the connection between concepts could

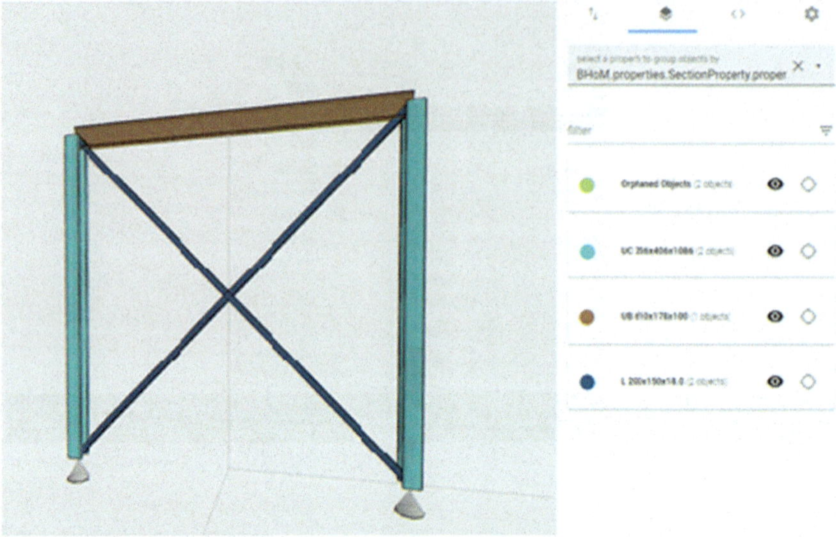

Fig. 7 A BHoM portal frame structure sent and visualised in Speckle v1.0

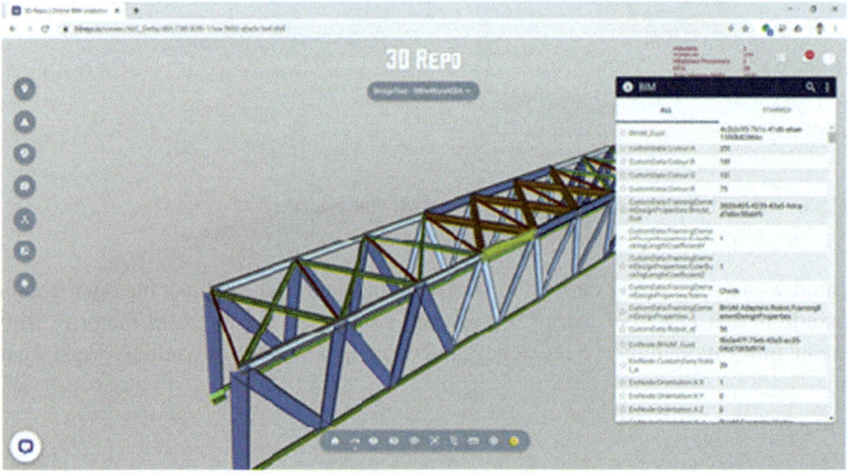

Fig. 8 A BHoM bridge structure visualised in 3D repo

allow for inferencing and reasoning on data. This is where the implementation of ontologies in the AEC industry could shine.

IFC already exposes IfcOWL, which provides a Web Ontology Language of IFC. There are conversion tools between IFC and RDF graphs (including IfcOWL) [44]. However, the potential is limited to what IFC can represent; therefore, IfcOWL inherits the limitations of IFC, like complex dependencies and relationships derived from the deep inheritance in the data structures, the limited dynamicity of the schema, and closed-source nature.

Looking at the open-source, grassroots schema of Speckle and BHoM, some observations can be made. Speckle mainly focuses on connecting software rather than representing building elements, and has a limited core schema, therefore the data is mostly represented via the schemas of the software that are being connected. In BHoM, there is a rich schema that owns several concepts in the built environment and alternative representations of the same concept seen from different domain lenses. For example, a Column is differently represented in a Physical oM (a schema namespace that represents tangible objects, like built elements) and in Structure oM (which represents objects as seen from a Structural Engineering view), where it is called Bar. Furthermore, due to how the classes are defined in the Object Models, and to the stringent rules that keep functionality and schema apart and maintain consistency across the framework classes, it is possible to find an equivalence between BHoM classes and RDF classes. In this sense, BHoM classes effectively can represent ontological concepts (ontological classes). Each class can be linked via other classes and to properties with predicates, and the same can be said of Engine methods, which always clearly target a main subject and return a clear object. In this sense, it becomes possible to map BHoM objects to Knowledge Bases via an ontological approach (Fig. 9). This concept is explored in the RDF_Prototypes toolkit, an open source collaboration between BH and the University of Stuttgart. [45]

The introduction of ontologies to BHoM is currently part of a research project from the University of Stuttgart, and it is exploring the potential of the BHoM framework to increase the use of knowledge graphs in the AEC industry, improve data interoperability and assist design decisions through inferential reasoning [46]. To this end, a specific BHoM Toolkit was developed that is able to convert BHoM objects to Turtle (Terse RDF Triple Language or TTL, a common format for ontologies) and export and visualise BHoM ontologies in WebVOWL, an ontology web visualizer [44] (Fig. 10).

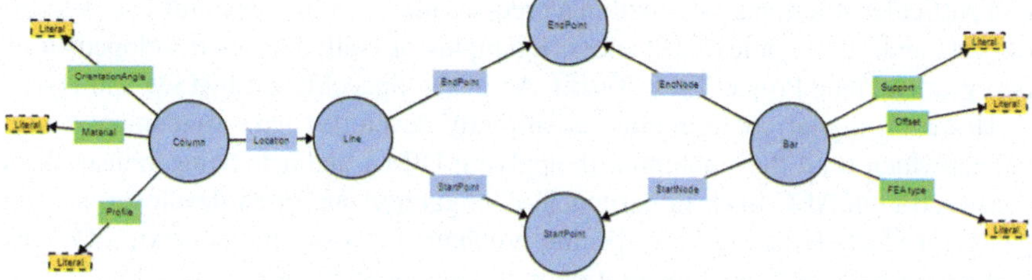

Fig. 9 Connecting BHoM concepts with an ontology approach: example with Column and Bar

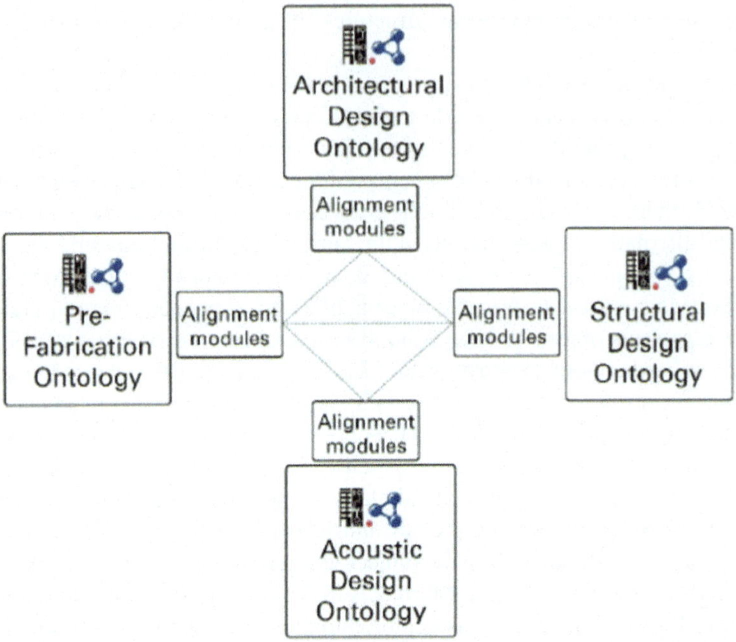

Fig. 10 Modular disciplinary ontologies using BHoM and semantic web technologies

4.2 *Configurators in AEC: The Need for a Common Framework*

The wider AEC industry has recently seen a growing number of platform-based approaches, focusing on the deployment of DfMA (Design for Manufacture and Assembly) and industrialized construction. A common aid to this approach is seen in the digital configurator, a software platform that allows for users to explore options of a particular solution space, within a predetermined set of constraints, such as kits of parts and other contextual factors. Examples of configurators developed in the past years include Project Frog, Testfit, Archistar, SpaceMaker, PRISM and Hypar.

However, digital configurators as currently developed have fundamental limitations which limit their potential in applying DfMA and achieving Industrialized construction in AEC [47]. In particular, configurators are often developed as standalone products targeting very specific problems for a specific context, a fact that limits their reusability and integration with other configurators. Furthermore, most digital configurators lack a capability for both downstream and upstream flow of information between the design and construction processes of a building, i.e., the upstream definition of design specifications against the downstream development of construction parts, components, and systems.

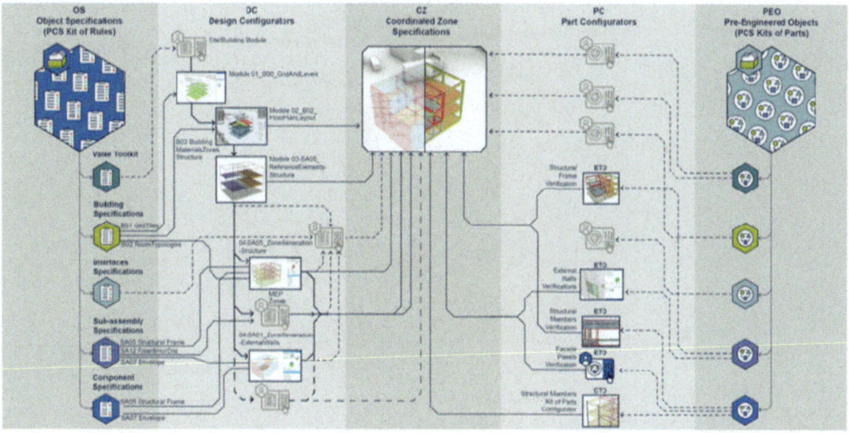

Fig. 11 Diagrammatic summary of the reference implementation with the common configurator framework

In 2021, a research project explored an alternative approach to unify digital configurators under a single framework. This research was developed under the Construction Innovation Hub (CIH) [48] and the Platform Design Program, a UK government funded initiative targeting successful DfMA. The proposed solution defines a Common Configurator Framework that institutes a separation between the authoring of objects for the design processes and the rules applied in the verification processes. The work demonstrated that this separation allows for a standardization and reuse of sets of parts (objects) and sets of rules (specifications). Additionally, the work formalized the decoupling of the downstream part verification process and the upstream design requirements definition, in order to allow for a Common Specification-driven design and manufacturing process. This was achieved by providing a mechanism for authoring and sharing objects and specifications, therefore enabling the intercommunication of multiple discrete configurators. The work was supported by a reference implementation which adopts the Open-source Framework and utilizes standard industry technologies, to demonstrate a series of case studies with prototypical configurators and examples of verification/validation targeting pre-existing kits of parts against pre-authored specifications. Future work is needed to identify and implement more case studies that explore the relationships between human-driven and machine-automated Validation and Verification processes [49] (Figs. 11 and 12).

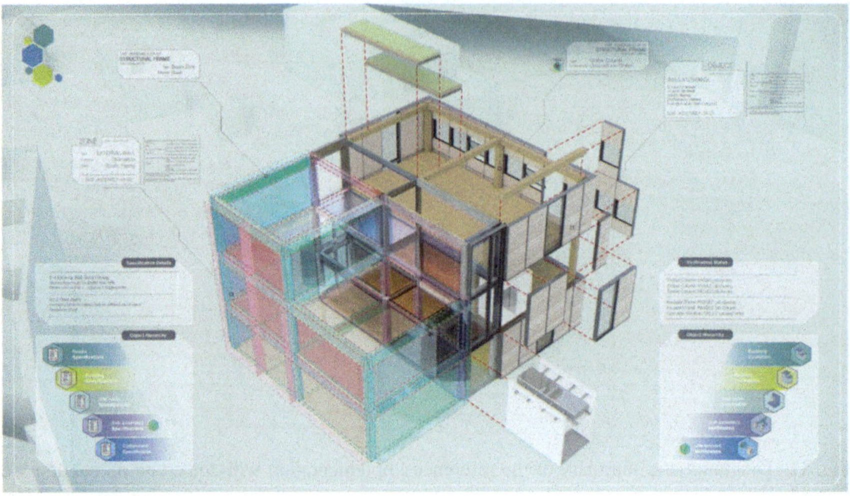

Fig. 12 Reference implementation: example of coordinated design obtained through several integrated configurators

References

1. Haymaker JR, Fischer CKM (2005) A methodology to plan, communicate and control multidisciplinary design processes
2. Werbrouck J, Pauwels P, Beetz J, Mannens E (2021) Data patterns for the organisation of federated linked building data
3. Elshani D, Wortmann T, Staab S (2022) Towards better co-design with disciplinary ontologies: review and evaluation of data interoperability in the AEC industry
4. Hummingbird. https://www.food4rhino.com/en/app/hummingbird. Accessed 17 Dec 2022
5. International Organization for Standardization (ISO). ISO 16739: industry foundation classes (IFC) for data sharing in the construction and facility management industries
6. buildingSMART (n.d.) Industry foundation classes. bSI Standards. http://www.buildingsmart.com
7. Bourreau P, Charbel N, Werbrouck J, Senthilvel M, Pauwels P, Beetz J (220) Multiple inheritance for a modular BIM
8. Venugopal M, Eastman CM, Teizer J (2015) An ontology-based analysis of the industry foundation class schema for building information model exchanges. Adv Eng Inform 29(4):940–957. https://doi.org/10.1016/j.aei.2015.09.006
9. https://www.w3.org/standards/semanticweb/. Accessed 19 Dec 2022
10. Hitzler P, Krötzsch M, Parsia B, Patel-Schneider PF, Rudolph S (2021) OWL 2 web ontology language primer. W3C Recommendation. https://www.w3.org/TR/owl2-primer/
11. Pauwels P, Terkaj W (2016) EXPRESS to OWL for construction industry: Towards a recommendable and usable ifcOWL ontology. Autom Constr 63:100–133. https://doi.org/10.1016/j.autcon.2015.12.003
12. https://w3c-lbd-cg.github.io/bot/
13. Esser S, Borrmann A (2021) A system architecture ensuring consistency among distributed, heterogeneous information models for civil infrastructure projects
14. https://speckle.community/t/introducing-kits-2-0/710. Accessed 20 Dec 2022
15. https://martinfowler.com/eaaCatalog/dataTransferObject.html
16. https://speckle.guide/dev/base.html#direct-usage

17. https://stackoverflow.com/questions/11889602/difference-between-strong-vs-static-typing-and-weak-vs-dynamic-typing
18. https://speckle.guide/dev/objects.html#specific-host-application-support
19. https://git-scm.com/book/en/v2/Git-Branching-Branches-in-a-Nutshell
20. https://speckle.guide/dev/server-webhooks.html
21. https://speckle.guide/dev/apps-auth.html
22. https://speckle.guide/dev/transports-dev.html. Accessed 20 Dec 2022
23. https://speckle.guide/dev/server-graphql-api.html
24. https://speckle.systems/features/connectors/
25. https://speckle.guide/user/support-tables.html
26. https://speckle.guide/user/revit.html
27. https://github.com/specklesystems/speckle-sharp/tree/main/Objects/Objects
28. https://github.com/specklesystems/speckle-sharp/tree/main/Objects/Converters
29. For example, https://github.com/specklesystems/speckle-sharp/tree/main/ConnectorRevit
30. For example, https://github.com/specklesystems/speckle-sharp/issues/1694 and https://github.com/specklesystems/speckle-sharp/issues/1942
31. https://github.com/specklesystems/speckle-unreal
32. https://speckle.guide/dev/architecture.html
33. https://speckle.guide/dev/architecture.html#kits
34. https://github.com/BHoM/BHoM/tree/main/Geometry_oM
35. https://github.com/BHoM/BHoM/tree/main/Structure_oM
36. https://github.com/BHoM/BHoM/tree/main/Architecture_oM
37. https://www.silasreinagel.com/blog/2018/11/12/using-orms-and-dtos-elegantly/
38. This can be done via the IImmutable interface. See more on https://bhom.xyz/documentation/Contributing/Development%20FAQ/The-IImmutable-interface/
39. https://bhom.xyz/documentation/Versioning/#decentralisation-of-the-upgrade-information
40. https://learn.microsoft.com/en-us/dotnet/csharp/programming-guide/classes-and-structs/extension-methods
41. https://github.com/BHoM/BHoM/blob/main/Structure_oM/Elements/Bar.cs
42. https://github.com/BHoM/BHoM_Engine/blob/main/
43. https://github.com/aecdeltas/aec-deltas-spec
44. Elshani D, Lombardi A, et al (2022) Knowledge graphs for multidisciplinary co-design: introducing RDF to BHoM. LDAC
45. https://github.com/BHoM/RDF_Prototypes
46. https://github.com/pipauwel/IFCtoRDF
47. Al Fisher, Tsakiridis G, Lombardi A, et al. A Common configurator framework for distributed design, collaboration and verification across the full AEC supply chain. https://doi.org/10.1088/1755-1315/1101/9/092042
48. https://constructioninnovationhub.org.uk/
enlargethispage-12pt
49. Elshani D, Wortmann T, Staab S (2022) Towards better co-design with disciplinary ontologies: review and evaluation of data interoperability in the AEC industry. LDAC
50. Elshani D, Lombardi A et al (2022) Inferential reasoning in co-design using semantic web standards alongside BHoM. Forum Bauinformatik, München

Alessio Lombardi is a software developer specializing in the AEC industry with over seven years of experience. With an MEng/MArch degree from the University of Pisa (Italy) and an MSc in Computer and Data Science at Birkbeck University (London, UK), Alessio currently is a Senior R&D Software Engineer (ML and Interoperability Lead) at Buro Happold (BH) in London, UK. His expertise lies in Software Engineering and Machine Learning to develop tools that enhance productivity, collaboration and interoperability for engineers and designers. Alessio's notable contributions include advancing the BHoM Framework and developing a configurator framework for the AEC industry, earning him prestigious awards. Alessio is now leading BH's efforts in the AI/ML R&D area. He actively collaborates with universities, including the University of Stuttgart (Germany) and University College London, on research and teaching activities.

Chapter 4
HAL Robotics: From Automated Architectural Design for Fabrication, to Architecting Automated Manufacturing Systems

Thibault Schwartz

Abstract This chapter chronologically presents the genesis of industrial robotics software specialists HAL Robotics. The first part focuses on the founding team's preliminary motivations and experiments with generative design for manufacturing, leading to the emergence of early robot simulation and control tools within the scope of architectural practice and academia (2010–2014). The second part follows with the design requirements and development milestones of the HAL Robotics Framework for adaptive industrial manufacturing (2014–2022). Typical deployment architectures of the Framework are discussed and illustrated by a series of case studies, ranging from complex toolpath generation for component prefabrication, embedded path planning for mobile construction robots synchronised with BIM models, to cloud-based factory orchestration for distributed manufacturing. The chapter concludes with a discussion on how architectural design thinking helped completing these complex cross-practice software engineering projects, and how variable automation technologies primarily designed for non-standard construction-component manufacturing have progressively percolated to other industrial sectors.

Keywords Mass-customization · Non-standard construction · Robotic fabrication

1 (2010–2014) HAL: Robot Programming, Simulation and Control for Grasshopper

HAL Robotics is a London-based software company focusing on novel applications of robotics in manufacturing, construction and creative industries, and specialised in the automation of complex and variable processes. Despite the very technical

T. Schwartz (✉)
HAL Robotics, Pill Box, 115 Coventry Road, London E2 6GG, GB, UK
e-mail: t.schwartz@hal-robotics.com

P. Ruttico (ed.), *Coding Architecture*, Digital Innovations in Architecture, Engineering and Construction, https://doi.org/10.1007/978-3-031-47913-7_4

character of this activity, we started with an uncommon profile, as a small group of architects and robotics enthusiasts, combining our knowledge and previous experience with machines and software design with the goal to change the way in which one can interact with industrial robots. Years of research and development ultimately led to the development of a unique technical solution to model, program, simulate, control and monitor industrial robots: the HAL Robotics Framework (2).

This first part discusses the preliminary work of HAL Robotics' founder Thibault Schwartz (*I* herein), spanning from his initial motivations and experiments with robots as an architecture student, to the release of HAL for Grasshopper, and the production of multiple demonstrators and training workshops during which he met HAL Robotics co-founders Tristan Gobin, Sebastian Andraos and Rémi Vriet.

2 The Challenges of Modelling and Manufacturing Curvature

Studying architecture at ENSA Paris-Malaquais in the late 2000s–early 2010s, I observed with eager eyes the deep changes in teaching methodologies and research trends burgeoning in a variety of foreign universities. These bleeding-edge studios had seemingly unlimited access to powerful computers and software, and an impressive collection of state-of-the-art machines to experiment with digital fabrication such as CNC routers, laser cutters or SLA 3D printers. After the release of Rhino 4 (2006) and the publication of the RhinoScript primer by David Rutten (2007), I started automating simple 3D modelling tasks during the end of my first year, getting a basic understanding of programming and what I would come to know was "generative design".

Not having access to a FabLab and guessing—correctly—that no CNC machine would be available at my school before the end of my studies, I taught myself software skills to emulate projects published by RhinoScript power users. My favourite blog was Francois Roche's and Marc Fornes' (n)certainties—Columbia studio [9], which proposed a truly unique aesthetic and technical display, radically contrasting with my traditional academic environment. A few months into my second year, Andrew Payne's Grasshopper primer was published, and I immediately adopted this tool to automate 3D modelling. The unparalleled ease of use of Grasshopper and the exponential growth of user and developer communities that followed its release allowed me to dive deep into parametric and generative design, starting a long series of modelling experiments from spatial lattice structures and shells, to entire projects, and ultimately automating the fabrication of demonstrators… with robots.

Despite my interest for the aforementioned research and the trend of complicated, blob-covered—though sometimes visually appealing—virtual models produced at the time, I personally tried using computers to help assess structurally- and economically-sound designs. This took an entirely different turn when I tried to manufacture physical models of these geometrically complex projects, which proved immensely difficult without access to CNC machines. Not only was it an issue for me as a student, but I started to understand that fabricating non-standard components was still an open issue for the construction industry, and that engineering teams working on flagship projects such as Frank Gehry's Fondation Louis Vuitton (FLV) or Zaha Hadid's Chanel Pavilion in Paris were struggling to materialise their designs despite a more or less unlimited budget. These teams faced many issues, most of them I would not pretend to grasp, from ill-adapted planning laws to the transportation of precisely machined van-sized facade elements, but I found it difficult not to see avoidable design and technical issues seemingly linked to modelling methodologies, gaps in software, and even choices in the manufacturing processes of curved components.

My graduation slowly approaching, I looked for a 4th year internship in architectural consulting groups dealing with complex modelling and fabrication topics and was lucky to spend a few months at Gehry Technologies Paris (GT) on the FLV site in 2011. This unique experience gave me the opportunity to witness the extreme complexity and technicity of such unconventional buildings, and to confirm my suspicions of software-related issues.

The FLV 3D model was a massive 120 GB versioned Digital Project (CATIA v5) model, shared and edited by up to 400 architects, engineers and subcontractors. The sheer effort required by GT to simply maintain the model's coherence was palpable: file revision systems were a novelty in AEC, and most participants were new to 3D modelling tools let alone using CATIA. The complete geometric complexity of the project—only floors were not curved—forced every actor to shift their workflow to a BIM-centric one, and new custom software utilities were constantly required to automate modelling steps, minimise human error and enforce model quality. Despite all these efforts, issues remained when the main model information had to be translated to manufacturing documentation. The steel industry was already adopting Tekla so the translation of thousands of metallic node assemblies and beams from one BIM standard to another and then to CAM was technically possible, but on the other side of the spectrum concrete fabricators were still exclusively relying on 2D shop drawings to manually manufacture formworks. Therefore, the modelling and fabrication of the 19,000 unique double-curved Ultra-High-Performance Concrete (UHPC) facade panels of the building proved to be particularly difficult.

The first challenge encountered came from the modelling technique used by the architect, Gehry Partners. Frank Gehry is famous for his architectural sketching and physical modelling workflow during preliminary design phases, which revolves around cardboard models digitised using 3D-scanning, used as a basis for the project 3D model. In theory these physical models can only be composed of ruled surfaces, which despite their curvature remain fairly simple to materialise as they can be discretised along rulings. Unfortunately, the 3D scans of these models are tessellated from point clouds, thus requiring rationalisation steps and surface fitting to attempt to rebuild the initial ruled surface. As this process is approximative, discrepancies between the expected surface properties and their digital representation are introduced. At the software level, these errors prevent surface modelling algorithms to

Fig. 1 The Fondation Louis Vuitton in Paris features a concrete skin of 19,000 curved UHPC panels cast in deformable moulds, vacuum-pressed against milled counter-forms ("MSV" process) [2]

function properly due to precision issues, impacting the ability to obtain connective seams between cladding components, introducing loss of planarity or unwanted curvature. At the building scale, this translates into panel fixturing, alignment and joining issues. At the prefabrication level, it complexifies formwork manufacturing and casting processes (Fig. 1) by forcing the manufacturer to use expensive techniques required for freeform panels. In theory, as we will discuss later, a single hot-wire cutting pass to obtain the same result would have been faster, more precise, and much cheaper.

The second challenge was a software workflow issue: no API was available to automatically program the CAM operations for the manufacturer's CNC machine, which forced operators to manually process thousands of machine jobs manually using a poorly designed vendor-specific CAM package, each of them taking at least 20 min to prepare when a simple API access—or use of a different software stack to control the CNC—would have solved this with only a few dozen lines of code.

This internship ended before I could see the final evolutions of this process, and so I cannot comment on how remaining issues were solved. The FLV now stands complete in Paris, a brilliant display of architectural design, engineering and construction technique. But my feeling when leaving GT was one of deep frustration: I had the intuition that an alternative approach, leveraging generative modelling and robotics, could achieve better results, faster, at the fraction of the cost, if only a coherent software stack could be assembled.

Mere days later, I started working on a set of software and hardware tools to demonstrate my hypothesis: that hot-wire cutting curved foam panels with a 6-axis industrial robot could be fully automated from modelling to fabrication, with no loss of precision, and for any number of unique parts.

3 (2011–2012) HAL V0.00x, Habitat Prospecteur

In 2010, a year preceding my internship at GT, I was invited by Philippe Morel and Felix Agid to do an internship at EZCT Architecture and Design Research, the architectural practice, experimentation and teaching platform they created with Jelle Ferringa (Odico, Aectual) in Paris in 2000. It was the perfect environment for me to apply and build-up my computational design skills, and our collaboration between 2010 and 2015 would lead to a variety of projects, some of which are presented below.

EZCT's acquisition of a small ABB IRB120 industrial robot in July 2010 gave me access to this technology a few years before it was introduced to the vast majority of architecture schools, hence constraining me to self-train and build my own hardware and software tools. I anticipated it was going to be quite a challenge and I knew it would take more time and skill than I had to get it to work, so the robot stayed in its crate for a few months. Coming back from the GT internship, I finally had found the "simple enough" kickstart project that would articulate gracefully with my own explorations and studies, and keep me motivated to get through the dozens of hours necessary to digest and implement the robot manuals, learn the manufacturer controller administration software (RobotStudio) and the extra CAM package that had been purchased with the machine (MasterCAM + RobotMaster). But in my mind, the goal was clear: I would "simply" run some quick form-finding to have a shell model, take a portion of it that I would approximate with ruled surfaces, discretise the strips to match the scale of the robot, and hot-wire cut moulds from the negative of each panel. All these things can be done today in less than a day of work, but the situation was quite different then.

After spending a few days and nights struggling on software compatibility issues and getting a simple toolpath to be uploaded to the controller, I realised the robotics and CAM software at my disposal were not going to be usable for what I had in mind: the proposed workflow was extremely cumbersome, revolving mostly around virtually jogging the machine using a mouse in the place of the teach-pendant joystick. I needed to be able to program tens of thousands of robot positions precisely to cut curved surfaces, so manual teaching was not an option. The CAM package was interesting, but required lots of manual input, took time to validate trajectories, and was essentially designed to convert milling toolpaths to robot procedures. Additionally, unlocking 5-axis machining strategies that I needed for my application was an optional upgrade which doubled the software's price. I was struggling to get the hot-wire cutting process to work, and having to disguise my cutter as a milling head was fundamentally counter-intuitive as the cutting direction was perpendicular to the simulated one. All things considered; the number of hacks involved to follow a 3D spline with the robot frustrated me greatly. I finished that week wanting to be free to implement any toolpath for any end-effector or process, and to do it directly from Grasshopper instead of having to convert my files into different formats, and re-processing them manually every time I needed to change a parameter.

Browsing the Grasshopper forum, I realised that while I was busy working out how these things could work, other members of the community had found themselves in similar situations only a few weeks earlier: Daniel Piker posted a geometric Inverse Kinematics (IK) definition for simple spherical-wrist robots called "Lobster" [12], and Johannes Braumann and Sigrid Brell-Cokcan had published a paper presenting a soon-to-be-released plugin similar to the one I was planning to implement but for KUKA robots [3]. While neither of these tools were exactly what I needed, it gave me proof that I was going in the right direction. A few days later, I had a working prototype of an ABB RAPID generator in Grasshopper with a preview of the robot position. The code was running on the controller, so I started prototyping moulds and compiled the definition into a set of Grasshopper components released mid-October 2011: HAL | Robot Programming and Control was born [13].

My theory was working as expected but casting concrete proved impractical and expensive to experiment with. Instead of cutting moulds, I switched to cutting curved blocks directly from foam. This enabled me to demonstrate the continuous generative modelling-to-fabrication workflow I wanted to obtain, fine-tune the toolpath strategies, demonstrate the cutting process, and assemble the results into a series of small pavilions between October 2011 and June 2012. Revisions of the HAL plugin were released in parallel to my experiments, opening the Robots in Architecture community—founded by Brell-Cokcan and Braumann in the wake of the KUKA|prc release—to ABB robot users.

My graduation project, Habitat Prospecteur (Fig. 2), named in homage to Jean Prouvé's project of prefabricated mobile cabins for petrol prospectors in Sahara, is a synthesis of all these experiments: a complete generative micro-house project with multi-objective optimisation mixing design and fabrication criteria with in-the-loop robot toolpath generation. Criteria included maximum volume, structural performance, estimated prefabrication time, robot trajectory feasibility, and a strict $1000 budget for the bill of materials.

4 (2011–2014) HAL for Grasshopper: A Robotics Training and Experimentation Toolkit for Architects, Designers and Engineers

Following my graduation and the nascent robotics trend within architecture and design schools, I started teaching computational design and robotics workshops in France, Austria, and the United Kingdom, and helped a series of academic and commercial institutions to get involved in this burgeoning experimental field in Europe, North America and Asia. As I travelled and encountered a variety of processes and cell setups, I expanded HAL for Grasshopper to include compatibility with KUKA and Universal Robot machines to ensure I could teach in any circumstance using a homogeneous software environment, with the possibility to port existing files from one training environment to another. The large variety of student

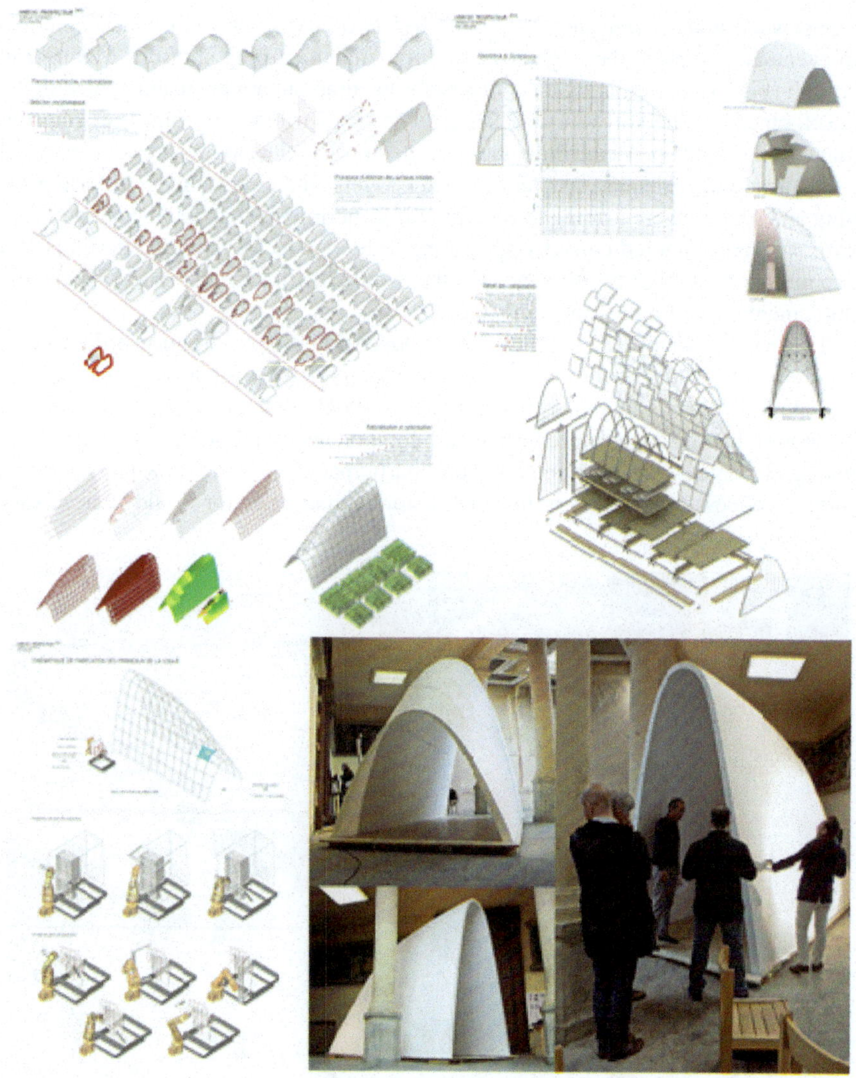

Fig. 2 The Habitat Prospecteur demonstrator aimed to prove the feasibility and pertinence of holistic computational models for construction

projects I had to supervise—up to 90 in a semester—was a good way to stress-test both the implementation and design of HAL for Grasshopper, as it had to comply with all sorts of applications for architectural and product prototyping, most of which I could not anticipate. These teaching activities led to a few interesting projects in the continuity of my research on lightweight masonry structures and economical non-standard manufacturing techniques, mostly at the UCL Bartlett B-Pro GAD where I introduced robotics between 2012 and 2014, assisting Philippe Morel and Guan

Lee (Figs. 3 and 4). Between 2013 and 2015 two HAL-powered projects emerged to eventually become the large-scale additive manufacturing start-ups MX3D and XtreeE (Fig. 5), highlighting an additional issue that had not been anticipated in the development of the software: the capacity to be distributed in industrial environments and scaled up. Meanwhile, the restricted project development cycles imposed by the academic calendar, and very limited budgets which prevented mid-term planning and scaling were leading to repetition in experiments that could be undertaken with students. I started considering leaving academia to focus on the development of a more robust robotics software suite that would facilitate technology transfer and implementation of industrial applications.

Having validated the HAL for Grasshopper's ability to produce offline robot programs efficiently from complex geometric models, my curiosity also gradually shifted towards human–robot interaction (HRI) systems. While offline robot programming is adapted to construction component prefabrication in a perfectly managed context, the experiments I had participated in were showing that some form of process adaptability or live interaction with an operator would be necessary

Fig. 3 Abeille vault prototyping workshop at UCL Bartlett, 2013 [14]. The production of small-scale demonstrators illustrating complex yet efficient structural morphology principles, robot programming, and assembly planning was at the core of the teaching the HAL Robotics team provided in a variety of architecture and design schools

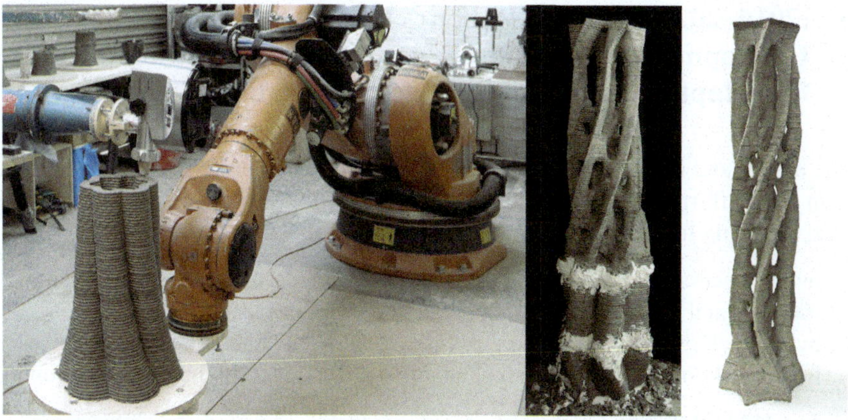

Fig. 4 Clay Robotics (J. Sun, K. Ho, S. Wang, tutors: Morel et al. [11]) demonstrated the realisation of cost-effective, recyclable 3D-printed clay moulds to cast complex concrete lattices [10, 11]

Fig. 5 The DEMOCRITE project allowed the production of UHPC walls and slab components, leading to the creation of 3DCP start-up XtreeE [6]

if robots were to be used in unstructured construction environments, or simply to compensate for unexpected process events. A series of HRI implementations using remote control interfaces and a variety of 2D and 3D sensors for object and gesture recognition were implemented from July 2012, gradually complexified and experimented with until 2016, following the creation of the HAL Robotics team [1, 15]. The technical complexity of these experiments led to refined design requirements for any future software, to ensure that any sensor and communication protocol could be handled gracefully.

Early in 2015, the HAL robotics team was assembled, I left academia and the development of the HAL Robotics Framework started.

5 (2014–2022) the HAL Robotics Framework: Development of a Vendor-Agnostic, Portable Middleware for Adaptive Industrial Robotics Applications

The design and implementation of the HAL Robotics Framework (HRF) is a long-term team effort [4]. As stated in the previous section, the creation of a new code base departing from the HAL for Grasshopper plugin was motivated.

- by the necessity to achieve industrial quality standards to facilitate technology transfers and the emergence of construction robotics businesses.
- by the lack of a reliable, coherent, easily deployable and maintainable software framework for the development of adaptive processes involving industrial robots.

The HAL Robotics Framework, is an extensible and modular software which enables adaptive programming of robot tasks, inter-device communication, and motion planning for one or many robots. It has the ability to reprogram robots automatically based on part or process variation. Additionally, it is vendor and process-agnostic, meaning that any process, robot or program can be simulated and translated to executable machine code. The software can be reconfigured to comply with any control and monitoring workflow, and is deployable on edge devices, web servers or workstations. A set of rich graphical user interfaces (GUI) can be configured to access the framework, either as a specialised standalone application, a web or mobile application, or as a CAD/BIM software plugin.

Major technical developments were undertaken to reach the desired level of flexibility, robustness and ease of use, including:

- a motion planning engine which supports multi-robot cells and management of kinematic redundancies, to be able to solve complex constrained motion sequences for any number of robots, positioners and AGVs working synchronously or asynchronously. This engine also enables motion planning of modular robot arms.
- an extensible procedure post-processing engine, allowing analysis and translation of simulated tasks to any robot-specific programming language.
- a flexible communication protocol builder, ensuring compatibility with both open and proprietary industrial network standards. This allows the HRF to run on-premises or in the cloud with adequate latency for command, logging or orchestration purposes.
- extensive equipment preset catalogues containing a thousand robots and dozens of controller emulations from the past 25 years, to quickly build digital twins of any production cell.
- an extensible, simplified Grasshopper plugin offering a unified work environment to use any robot, and easily develop application prototypes.
- a standalone application development toolkit facilitating the development of customised UX for any user role or process.

Having such a modular and flexible set of software components allows the implementation of safe, easy-to-use adaptive robot programming, simulation and monitoring workflows running on a variety of devices for optimal performance. The openness of the system makes it the ideal platform for the scheduling of hybrid processes to be executed simultaneously or in sequence, and for just-in-time reprogramming of tasks based on sensor readings or other data sources.

6 CAD-Based Programming Workflow and Derivatives

The most typical HRF implementation workflow consists of detecting and associating CAD part features or metadata with algorithms generating trajectories to be executed by one or more robots (Fig. 6). Using a digital twin of the production cell, the trajectories are simulated to identify and prevent risks of collisions, reach or speed limitations. Once validated, the simulated procedures are translated to executable robot code adapted to the target machine controller model, and added to the task execution queue.

An extension of this workflow fetches execution data from the robot controller and/or process equipment for traceability purposes, or for tuning process parameters (Fig. 7). Simulated measurements can be used to evaluate cycle time, train machine learning (ML) models, or optimise product or cell design parameters (Fig. 8), as demonstrated in the following case study.

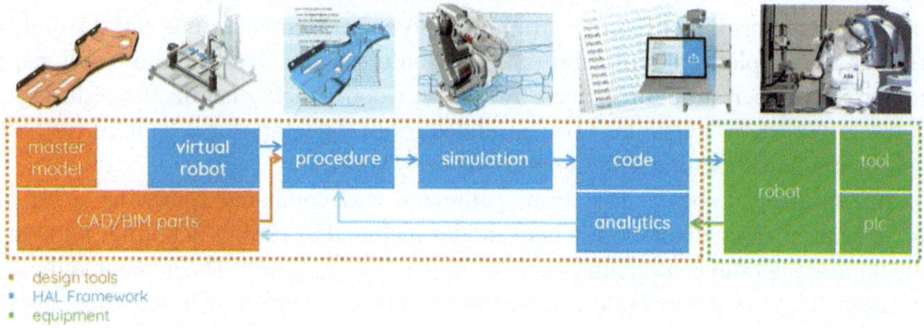

Fig. 6 Typical CAD-based offline programming workflow, using the HRF to translate product properties into robot instructions. Depending on the robot controller generation, execution data can be retrieved and looped back to modelling or programming steps

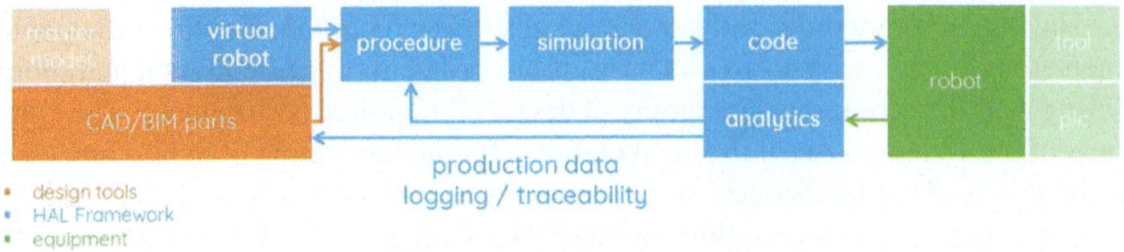

Fig. 7 Advanced offline-programming and monitoring workflow, storing task execution data back into the production model for traceability

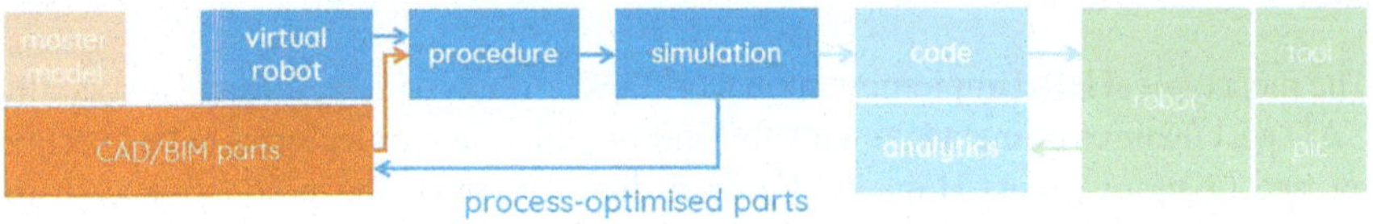

Fig. 8 Advanced offline-programming and product optimization workflow, using the robotic process simulation results as fitness criteria to optimise parts to be manufactured

7 (2017) Building Freeform: Integrated Design and Prefabrication of a Timber Shell-Nexorade Hybrid

Since 2015, HAL Robotics has been assisting the École des Ponts ParisTech in their research and prototyping of exotic structural systems in timber, composites and 3D-printed UHPC. As robotics software partner, we participated in the realisation of large-scale prototypes demonstrating Design for Manufacturing and Assembly (DfMA) principles and structural optimisation concepts. One of these pavilions, a reciprocal timber structure braced by flat plywood panels manufactured during a 1-week workshop, illustrates how simulated robot trajectories can be used for both cell layout and structural design optimisations [5] (Fig. 9).

In this project, beam and panel geometry was computed following a set of form-finding steps. Each component model included assembly detailing, allowing machining operations toolpaths to be computed automatically. The generated model was then used to validate both structural behaviour and machining trajectories. Being able to validate production feasibility for the entire project at every form-finding iteration allowed the design team to explore the project design space in an informed manner, and immediately start production when a variant was selected.

Fig. 9 The automated prefabrication of each beam of this complex structure involved handling, sawing, drilling and milling operations executed by a 13-axis multi-robot cell

8 (2018–2021) Autonomous Drilling and Fastening Robot Programmed from BIM Model

Although the previously demonstrated workflows are adapted to control robots in known production environments, a departure from this monolithic, continuous implementation has been required to allow local adaptation of a previously specified robot task. This can be achieved with the generation of intermediary job models to be stored, distributed and executed by machines when they become available after navigation and environment discovery steps, resulting in a simplified form of robot task orchestration (Fig. 10).

Since 2018, HAL Robotics has been working with a range of partners (Skanska, ABB, InnoTecUK, Skyjack, BRE) to develop mobile robots with the ability to navigate construction sites and achieve tasks sent to the robot from a BIM model. In its latest iteration, this Collaborative, On-Site Construction Robot (COSCR, Fig. 11) was deployed to autonomously and safely navigate around a construction site, and perform tasks that have been scheduled by the user—in this case, drilling and fastening anchors in concrete slabs from locations specified in the BIM model.

The solution was developed to specific size and weight constraints to maximise its potential to access the tightest corners of any site. The platform is tracked, which enhances manoeuvrability, and features a telescopic mast fitted with a robotic arm, enabling it to reach heights of up to 4.2 m.

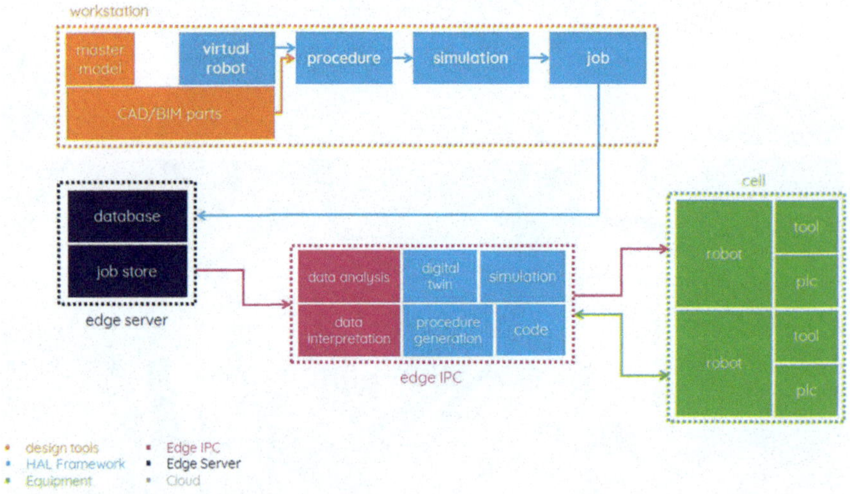

Fig. 10 Typical distributed job architecture for a hybrid mobile robot, where the HRF is used as a plugin in the CAD/BIM environment, and as an embedded client application in the robot to control a mobile platform and robot arm to perform a variety of tasks based on local measurements taken by sensors

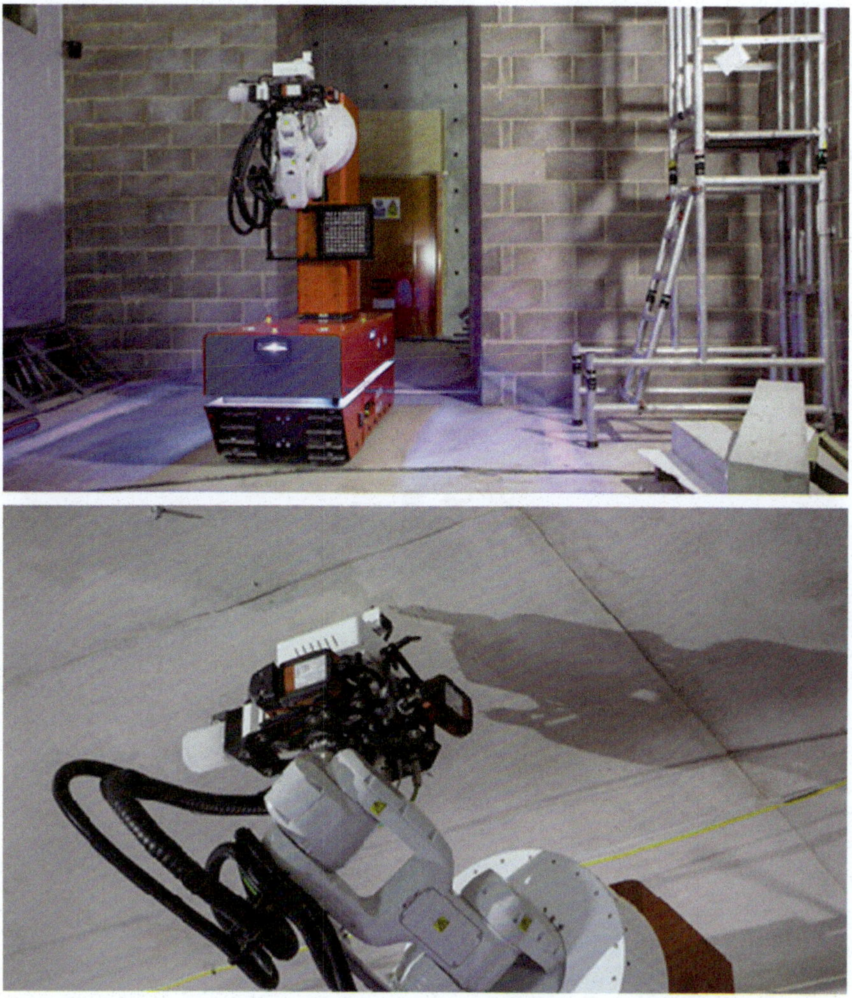

Fig. 11 The COSCR demonstrator deployed on site [7]. The robot was equipped with a triple end-effector for sensing, drilling and fastening at precise locations in slabs

9 (2022) Automated Programming of Large, Complex Assembly Inspection

A variation of this CAD-based job orchestration workflow has been implemented for the quality control of train bogies by AGV-mounted robot arms equipped with 2D and 3D cameras. The bogie models and metadata are converted, analysed and rationalised through a series of automated processing steps taking into account part taxonomy, generating inspection goals corresponding to assembly qualities to be measured such as dimensional tolerances, part presence, or compliance with other quality criteria.

Each goal is then converted as a set of visual acquisition steps to be completed with 3D or 2D cameras at specified focal lengths, definition and illumination settings (Fig. 12).

The elevated risk of collision and visual occlusion when positioning the sensor, associated with the large scale and geometric complexity of the assemblies, led to the necessity to develop path planning strategies mixing kinematic workspace analysis with process objectives, and integrate variable tolerance estimation to compensate for any imprecision of the AGV's localisation. Some of this imprecision was reduced by accurate measurement of the robot position relative to the inspected vehicle using large field 3D sensing, partially regenerating the scheduled robot job just before its

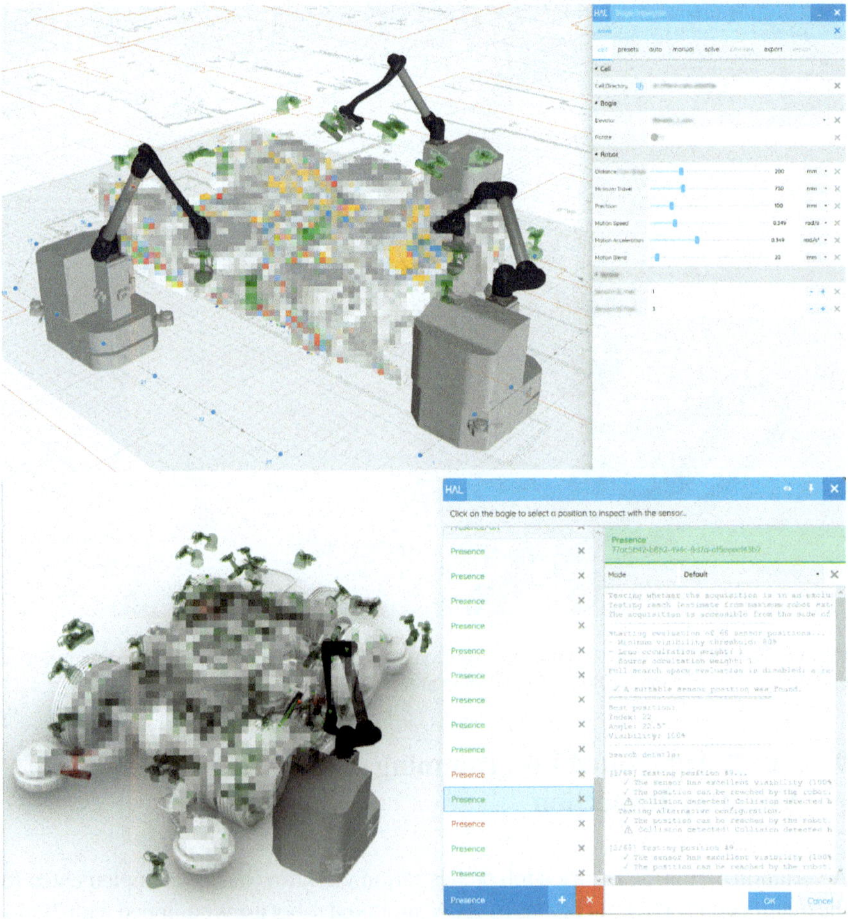

Fig. 12 The Train Bogie Inspector application is composed of a standalone software interface to generate and edit inspection campaigns on thousands of parts, and of a secondary on-board software solution responsible for compensating robot pose deviation, execute measurement tasks and generate inspection reports. The motion planner decisions are logged for traceability

execution. The digital twin was also used to preliminarily tune image and point cloud processing algorithms based on simulated sensor results.

10 (2022) Distributed Sensing and Robot Job Generation

Instead of being fed from a pre-established CAD model, the HRF can also be configured to link sensors with distributed storage and data processing services running on different machines at different latencies, eventually emitting robot jobs based on measurement interpretation. This loosely-coupled architecture offers the ability to use a variety of sensors and analysis techniques interchangeably with reasonable implementation costs. Additional work was also completed to balance network payload for such distributed architectures, in order to minimize cycle times [17] (Fig. 13).

This hybrid sensing and processing system is currently being explored to automatically detect defects on manufactured products before generating adapted countermeasures, with the aim to improve accuracy and quality of parts and minimize scrap. A first demonstrator implemented laser scanning of 3D-printed parts to detect dimensional deviations of contact surfaces, that would then be machined to reach desired tolerances (Fig. 14).

A variant of this measure and process workflow is being implemented to discover the geometry and location of parts without knowing their dimensions or features in advance, then process them. In this scenario, kitchen cabinet doors of various sizes and designs are presented to the machine to be sanded before painting or varnishing (Fig. 15). The part discovery replacing CAD-based programming is motivated by the extreme variability of product orders, and the lack of automated product handling system in the assembly workshop.

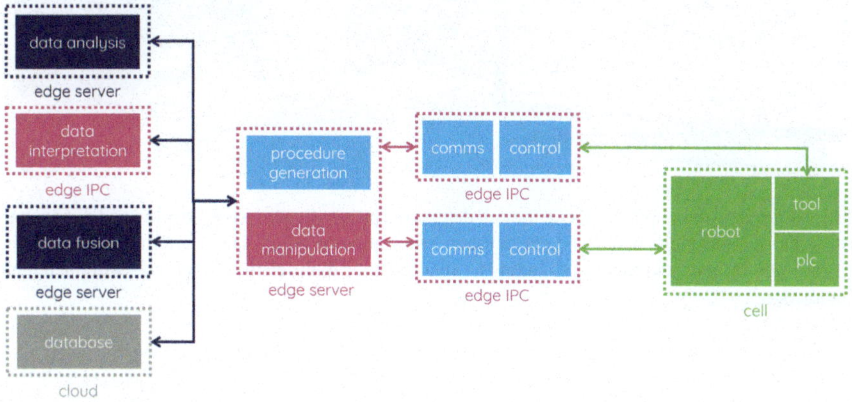

Fig. 13 Distributed architecture for adaptive control through independent micro-services for control, monitoring and analysis

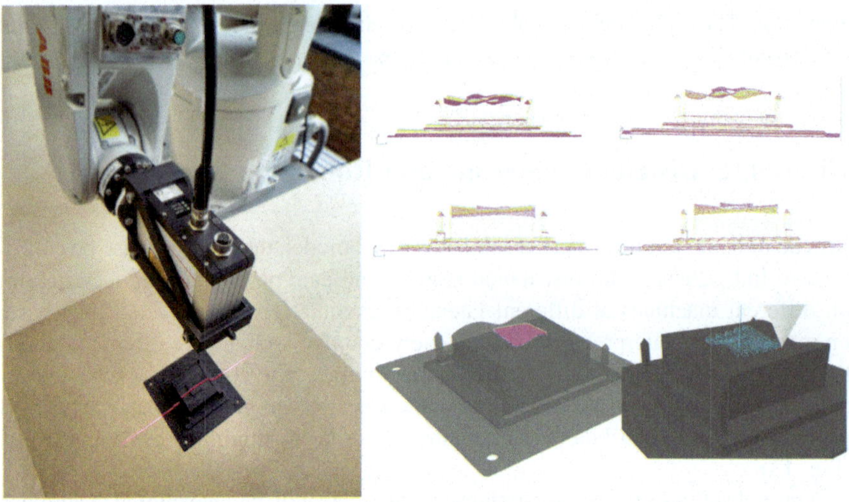

Fig. 14 Defect detection and correction using a robot-held laser sensor and a spindle. Both end-effectors are mounted on automatic tool changers so the robot can switch between inspection and processing jobs iteratively until the desired part quality is obtained

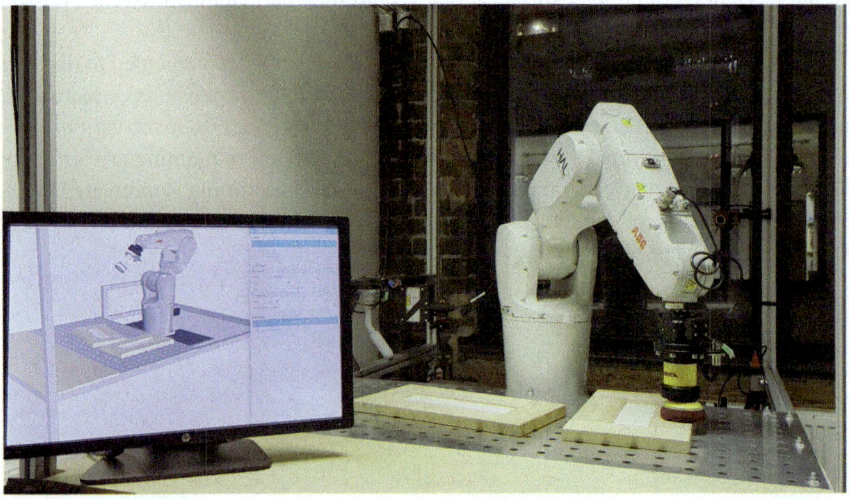

Fig. 15 A 3D scanner placed above the cell detects the presence, geometry, position and dimensions of kitchen cabinet elements placed near a robot. The scanned product is then automatically analysed to generate, validate and execute a sanding toolpath adapted to the element's features without human interaction

11 (2022) Cloud-Based Robot Job Validator and Orchestrator

As demonstrated in 2.3, 2.4 and 2.5, the HRF can be used to pass jobs to robots in a more or less distributed fashion based-on performance and execution constraints. A generalisation of this is the ability to orchestrate the execution of a pool of jobs, based on the identification of robot cell manufacturing capabilities [8]. The Distributed Manufacturing for Off-site Construction (DMOC, Fig. 16) project was a prototype, developed in partnership with Konica Minolta and aLL Design, aiming to validate a manufacturing model enhancing the DfMA approach discussed in 2.2 with just-in-time (JIT) distributed fabrication. In this system, production cells—"workers"—geographically disseminated at a variety of fabrication facilities, subscribed to a job matching and scheduling platform. After receiving production capability information from the worker, the job matcher compares production tasks published from a BIM model, and simulates the execution of the fabrication process using the worker digital twin. When the best available worker cell is identified, the job is transferred to it, translated to machine code and executed by the machine.

Building on this, another tool was developed on Amazon AWS to demonstrate the pertinence of cloud-based robot job aggregation and validation using containerised HRF instances simulating different processing strategies and settings. The ability to run massively parallel processing on-demand allowed identification of adapted

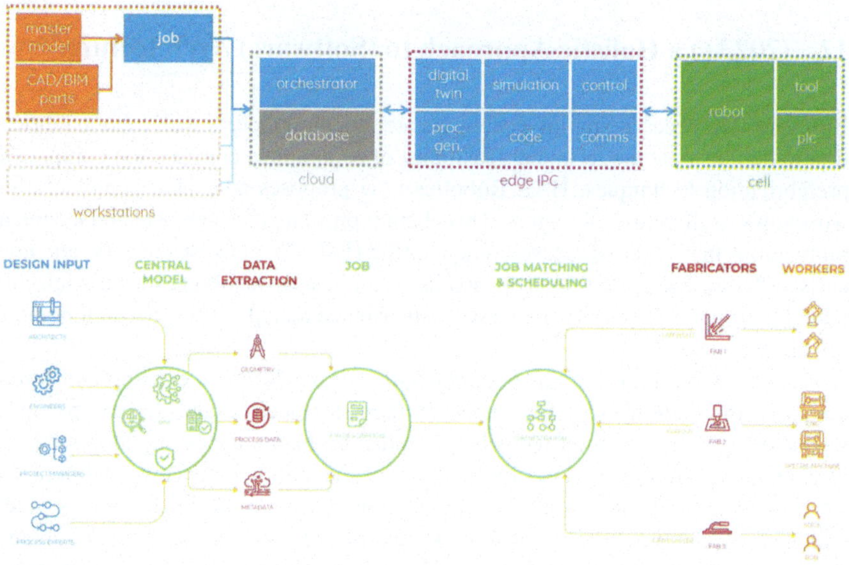

Fig. 16 The DMOC project allows fabrication jobs to be pooled and dispatched to distributed fabrication facilities based on production cell specifications and availability

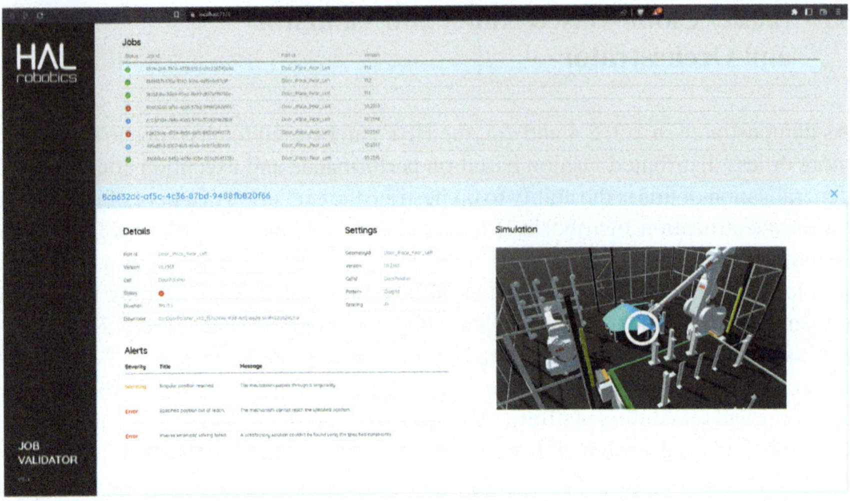

Fig. 17 A preview of the web orchestrator used to demonstrate adaptive reprogramming of industrial robots and toolpath validation in the cloud [16]

strategies in seconds, an acceptable processing delay in a production environment (Fig. 17).

12 (2023+) a Holistic Approach to (Software) Architecture

Born from the necessity to link generative architectural models with robotic fabrication to demonstrate the technical feasibility of economical construction component prefabrication techniques, HAL Robotics now proposes adaptable manufacturing workflows to all industrial sectors. This chapter presented the personal and technical motivations that lead to the development of the HAL Robotics Framework, and how early software design decisions opened large application landscapes, from high-mix manufacturing to non-destructive testing, distributed manufacturing and autonomous on-site robotics.

As a team of architects turned roboticists, we would invite existing and future actors of the building sector to consider their engineering and design speciality in the larger scope of industry, where similar problems are being addressed but rarely shared. Architectural design thinking, or the ability to conceive and organise systems of systems and to identify innovation spaces at the interaction between technical specialities, was of tremendous help in our quest for the ultimate robotics software platform. Architects, as generalists, can thrive in these technological interstices where—perhaps counter-intuitively—regulations and industrial structures can be more accommodating. Living at the intersection of process engineering, programming and product design, we like to think of ourselves as twenty-first century master builders, only nowadays Cathedrals are in code, and can change shape every week. Our hope is that architects and builders, confronted to the combined economical, ecological and sociological challenges of today, are now ready to participate in the current industrial and technological reshuffle and embrace creative automation.

After all, we made sure they can speak to robots in any way they want but it's up to them to continue the conversation.

References

1. Andraos S (2015) DMR: a semantic robot control language. In: Proceedings of the 33rd eCAADe conference, vol 2. Vienna University of Technology, Vienna
2. Aubry S, Bompas P, Vaudeville B, Corvez D, Lagrange T, Mazzacane P, Brizou A (2013) A UHPFRC cladding challenge: the fondation Louis Vuitton pour la création "Iceberg". In: RILEM-fib-AFGC international symposium on ultra-high performance fibre-reinforced concrete 2013, pp 37–48
3. Braumann J, Brell-Cokcan S (2011) Parametric robot control: integrated CAD/CAM for architectural design. In: ACADIA 11: integration through computation. Banff, pp 242–251
4. Gobin T, Andraos S, Schwartz T, Vriet R (2022) A framework for modelling, simulating and parametrically programming heterogeneous industrial machines. Constr Robot 6(2):1–8
5. Gobin T, Mesnil R, Douthe C, Margerit P, Ducoulombier N, Demont L, Caron J-F (2018) Form finding of nexorades using the translations method. In: ROBARCH 2018: robotic fabrication in architecture, art and design 2018. Springer, Cham, pp 232–241
6. Gosselin C, Duballet R, Roux P, Gaudillere N, Dirrenberger J, Morel P (2016) Large-scale 3D printing of ultra-high performance concrete—a new processing route for architects and builders. Mater Des 100:102–109
7. HAL Robotics (2020) COSCR. Retrieved from HAL Robotics: https://hal-robotics.com/use-cases/coscr/
8. HAL Robotics (2022) DMOC: Distributed Manufacturing for Off-site Construction. Retrieved from YouTube: https://www.youtube.com/watch?v=ZfoKS_ESH_U
9. Hong K (2007) Woven Habitat_Robot. Retrieved from (n)certainties—Columbia. https://new-territories.com/000%20my%20web%20site/n1/ncertainties.wordpress.com/students/kyuseon-hong/kh_final/index.html
10. Morel P, Schwartz T (2015) Automated casting systems for spatial concrete lattices. In: Modelling behaviour—design modelling symposium 2015. Springer, Cham.
11. Morel P, Lee G, Schwartz T (2014) Applications of robotics to discrete and continuous spatial lattices. In: Migayrou F, Porter A (eds) March GAD 2013–2014. The Bartlett School of Architecture UCL, London, pp 74–91
12. Piker D (2011) Lobster. Retrieved from Grasshopper forum: https://www.grasshopper3d.com/group/lobster
13. Schwartz T (2012) HAL: extension of a visual programming language to support teaching and research on robotics applied to construction. In: Rob | Arch 2012. Springer, Vienna, pp 92–101
14. Schwartz T, Mondardini L (2014) Integrated design and robotized prototyping of Abeille's vaults. In: Robotic fabrication in architecture, art and design 2014. Springer, Cham, pp 199–209
15. Schwartz T, Andraos S, Nelson J, Knapp C, Arnold B (2016) Towards on-site collaborative robotics—voice control, co-speech gesture and context-specific object recognition via ad-hoc communication. In: Robotic fabrication in architecture, art and design 2016. Springer, Cham, pp 388–397
16. Wallace J, Andraos S (2022) Adaptive reprogramming of industrial robots and toolpath validation in the cloud. Retrieved from Amazon Web Services Robotics Blog. https://aws.amazon.com/blogs/robotics/adaptive-reprogramming-of-industrial-robots-and-toolpath-validation-in-the-cloud/

17. Wang Y, Friderikos V, Andraos S (2022) A minmax utilization algorithm for network traffic scheduling of industrial robots. In: ICC 2022-IEEE international conference on communications. IEEE, pp 2936–2941

Thibault Schwartz has been developing robot simulation software for the past ten years. Trained as an architect, he actively participated in the democratisation of robotics in AEC and creative industries. He has since passed his knowledge and passion for automation to dozens of researchers and professionals around the world, and stays committed to crafting the best tools robot users can think of.

Chapter 5
Resilience and the Metaverse: A Toolkit Approach

Carlo Beltracchi

Abstract This chapter delves into the intricate relationships between man and cities as complex systems, highlighting the need for resilience and ethical systemic evolution within urban spaces to confront environmental challenges. We discuss the environmental and ecological impacts of cities and underline the pressing need for rethinking urban planning models. With advancements in technology, we explore the role of computer science in understanding urban structures and the increasingly significant role of the Web-3 era and the emergent data economy in city modeling. We further elaborate on the potential of metaverses and the role of community interactions within these virtual spaces, proposing an innovative approach for post-disaster reconstruction where architectural theory intersects with the metaverse, artificial intelligence, and blockchain technologies. The research builds upon the theoretical foundation of Lebbeus Woods' reconstruction series and explores a novel metaverse-based model for post-disaster reconstruction. To demonstrate the practical application of the proposed model, we envision a case study illustrating its use in a disaster scenario. Following this, we discuss critical points and potential limitations of the model. The study concludes with the implication that leveraging technology and reorienting urban planning towards preparedness and resilience could radically reshape post-disaster reconstruction, potentially paving the way for more efficient, responsive, and sustainable cities in the future.

Keywords Metaverse · Urban resilience · Web-3 · AI · Planning theory

---

C. Beltracchi (✉)
Building Engineering Architecture, Politecnico Di Milano, Via Gaetano Previati 1C, 23900 Lecco, Italy
e-mail: carlo.beltracchi@polimi.it

P. Ruttico (ed.), *Coding Architecture*, Digital Innovations in Architecture, Engineering and Construction, https://doi.org/10.1007/978-3-031-47913-7_5

1 Introduction

1.1 *Man and City: Complex Systems*

Man and city are complex and interdependent systems that change and emerge over time through a continuous relationship between them and with the environment or more generally the nature that surrounds them.

Complex systems are made up of multiple interacting actors that generate collective behaviors and cannot be simply explained by the sum of the individual parts [1].

From an ecological point of view, the idea of "biodiversity" suggests that the presence of multiple species in an ecosystem can contribute to greater stability and resistance to disturbances [2].

It is therefore possible to trace an initial connection between the concepts of *differentiation* [3], *redundancy*, and *resilience.*

1.2 *Resilience as an Intrinsic Trait of Complex Natural Systems*

In thermodynamics, a thermal reservoir is a system with a large thermal capacity that can absorb or release heat without significantly altering its temperature [4]. The reservoir can be considered a resilient system by definition, capable of absorbing thermal disturbances without altering its state.

Drawing on this analogy, resilience can be seen as the ability of a system to absorb disturbances without significantly altering its structure or function. For example, a resilient ecological system could absorb the impact of an environmental disturbance, such as climate change, without suffering irreversible damage to its biodiversity and functionality.

However, it is important to consider that the resilience of the system does not necessarily mean that it does not suffer any impact from disturbances [5]. As in the case of the thermal reservoir where thermal energy is still stored in the reservoir but its temperature does not undergo a significant change, similarly, a resilient system can suffer some impact from the disturbance, but its structure and functionality remain intact.

Resilience is therefore an intrinsic trait of complex natural systems, and refers to their ability to adapt to disturbances and maintain their functionality.

As expressions of such systems, it is possible to assert that man and city are characterized by complexity, diversity, and unpredictability, but also by adaptability and resilience.

Today, the complexity of individuals and cities is increasing even more, thanks to the growing interconnection between people and communities, rapid technological developments, and the impact of human activities on the environment and ecosystems.

All this makes it even more important to understand complex systems and their resilience, to address current risks and ensure a sustainable future [6].

2 Cities as an Environmental Challenge

2.1 The Impact of Cities on the Environment and Ecosystems

Cities have become the main habitat for humanity: in 2018, it was estimated that 55.3% of the world's population lived in urban settlements.

By 2030, it is projected that urban areas will host 60% of people globally, and one in three people will live in cities with at least half a million inhabitants [7].

Being true "attractors", their centrality entails a significant influence on the environment and surrounding ecosystems.

Human activities in cities, such as energy consumption, waste production, air and water pollution, deforestation, and the destruction of natural habitats, contribute significantly to the global environmental crisis.

The main challenge for humans in the city is therefore to find a way to coexist sustainably with the natural environment, minimizing the negative impact of anthropogenic activities on the surrounding ecosystem [8].

2.2 Towards Systemic Evolution: The Ethics of the Suitable.

The key to addressing this challenge is not so much short- or medium-term human phenomena, identifiable as *revolutions*, but rather gradual and nonlinear processes of systemic *evolution*, in which humans not only use their wisdom, understood as the ability to understand and govern the laws of the world for what appears immediately, but also and above all exploit their own sagacity, derived from the contemplation of these laws over time and capable of positively welcoming their aspects of change and sometimes apparent imperfection.

Replacing the ethics of the optimal with the ethics of the suitable [9] means ensuring the sustainable use of resources and respect for the natural environment as a fundamental prerequisite, moving from a deterministic and mechanistic view of the world to an ecological-systemic view, in which human beings are an integral part of the global ecosystem.

Implementing such a change requires a deep understanding of complex systems, particularly the economic, productive, and social systems that underpin city life.

This understanding requires the support of appropriate tools to measure and assess the impact of human activities on the environment, as well as to identify opportunities and solutions to address their negative consequences in a timely manner.

3 Understanding the City

3.1 The Importance of Models

A first step in addressing the problem of understanding a complex system like the urban one is to trace its relevant aspects in the form of fundamental elements and relationships that can constitute a model. In general, it is useful to distinguish three fundamental categories of models [10]:

- **Iconic models**, where each component of a given reality is represented through a sign in a one-to-one correspondence with it, that is, where each sign corresponds to a given element of reality. These can be graphic representations or lists and are the basis of every other type of model.
- **Logical models,** in which the relationship between symbols is emphasized rather than the representation of the symbol—as in iconic models. Mathematical functions and, more generally, algorithms belong to this type of model.
- **Operational models,** which have the function of providing the most complete representation possible of reality or experiences perceived as very complex, for which the historical aspect, that is, uncontrolled and, therefore, generally unpredictable, is considered non-negligible.

Operational models typically translate into games that can be:

- **Role-based**: where each actor has a role.
- **System-based**: with total freedom of self-organization.

Once the fundamental notions related to models are introduced, it is necessary to focus on the strong gradient of innovation offered by the recent developments in computer science, with particular reference to computer graphics and networks, in the context of studying the city as a complex system whose resilience can be somehow monitored through a continuous and historically relevant profiling of salient data.

3.2 The Role of Computer Science

Computing in urban planning and management is increasingly important, thanks to the exponential improvement in computing power offered by GPU computing as well as emerging disciplines such as data science.

The smart city model, through the use of sensors and IoT devices, describes simultaneous communication protocols between the physical city and the virtual city and represents a constant source of significant data [11].

Moreover, thanks to Building Information Modeling, urban models can be represented in a semantically relevant and measurable way through software or simulation environments capable of extracting useful information [12].

Photogrammetry, particularly aerial photogrammetry using drones through "Lidar" systems, allows transforming or collecting datasets—whether images or point clouds—and thus creating realistic iconographic models that accurately reproduce urban spaces [13].

Furthermore, through object-oriented programming languages integrated into game engines, it is possible to create categories of actors and functions that define extremely complex algorithmic interrelationship logics.

This allows for the development of urban models that include social, economic, and environmental elements and can be used to assess the impact of various endogenous or exogenous components on the city. [14]

Finally, the latest developments in the field of networks and the internet, such as 5G technology and broadband, have made it possible to create online video games in which many players can simultaneously be part of the same server and thus provide live information to this new form of operational model [15].

4 The Function of the Model in the Web-3 Era

4.1 Towards a Data Economy—Internet of Value

In the previously described context, blockchain technology plays a cardinal role and can support the arrival of a new economic paradigm: the data economy. In this context, where the central or polycentric model of web 2.0 is overcome, a distributed network scheme with reward logics for network nodes is reached [16].

This scheme, whose amorphous structure resembles a rhizome, has the ability to ensure the resilience of the system from the point of view of both privacy and data ownership, through the application of the scalability-security-decentralization trilemma.

The application of the concept of data economy translates into an operational model in which humans and machines can contribute to data provisioning and continuous, real-time system reinforcement. For example, each connected user, whether human or machine, is linked to their own identity and wallet, and has the opportunity to share significant data with the system, receiving an evaluation and then a reward from it [17].

Blockchain, particularly when supported by artificial intelligence, thus represents an indispensable technology for the implementation of this operational model [18] in relation to cities, and especially smart cities.

Thanks to its ability to ensure the distribution and security of data, it can contribute to the creation of an ecosystem where data, services, and agents are interconnected efficiently and securely and incentivized by a base economy.

In this regard, it is appropriate to quote the words of American cryptographer Ralph Merkle, who, referring to the blockchain and particularly to Bitcoin, states:

> ***"Bitcoin is the first example of a new form of life."***
>
> It lives and breathes on the internet. It lives because it can pay people to keep it alive. It lives because it performs a useful service that people will pay it to perform. It lives because anyone, anywhere, can run a copy of its code. It lives because all the running copies are constantly talking to each other. It lives because if any one copy is corrupted it is discarded, quickly and without any fuss or muss. It lives because it is radically transparent: anyone can see its code and see exactly what it does.

These statements have contributed to the development of the Nature 2.0 concept by Trent McConaghy [19], an engineer and entrepreneur specializing in artificial intelligence and blockchain. The concept combines emerging those technologies with the conservation and sustainable use of natural resources.

The idea behind Nature 2.0 is to create a digital infrastructure that allows for more effective and sustainable management of our planet's resources, creating an integration between nature and technology.

There are applications of this concept in the field of architecture as well: an extremely interesting example is No1s1 [20], a research project aimed at creating the first large-scale Decentralized Autonomous Space (DAS) prototype.

It is a self-owned meditative space, focused on the interaction between community and physical space through intelligent sensors and devices (IoT) regulated by smart contracts.

The minimum viable prototype (MVP) aims to demonstrate the autonomous space concept, stimulating discussions on its socio-technological impact and identifying technical, legal, and regulatory challenges for future research.

4.2 *The Metaverse and the Importance of Communities*

The presented operating model, which not only represents objects, logics, and relationships but is a true virtual dual of the city [21] in a biunivocal and simultaneous correspondence with it and its participants, represents a general vision of what we might call a metaverse applied to the urban, economic, and social theme.

What emerges from this model is the potential ability to produce inclusivity and differentiation through an economy based on the tokenization of assets, or by attributing value to reality starting from the information contained in the reality itself. This entails the possibility of defining a historical chain of rewards and reputation capable of recognizing a role and consensus for each participant who has actively contributed to the structure of the created information, providing relevant data or significant logics.

5 Innovative Post-Disaster Reconstruction: Architectural Theory Meets the Metaverse, AI, and Blockchain

5.1 From Theoretical Concepts to Practical Applications

Taking the discourse forward from our exploration of the transformative potential of the data economy and metaverse in urban development, we now delve into a distinctive application of these concepts. The realm in focus is post-disaster reconstruction, and the lens through which we examine it is inspired by the architectural theories that emerged during the late 20th-century, a period marked by significant cultural and intellectual ferment.

This era saw visionaries such as Lebbeus Woods reshaping the relationship between built environments, human societies, and the dawn of new technologies. In this context, we bring to light the concepts of *heterarchy* and *cyclicity*, both of which challenge traditional hierarchies and linear development models.

Bearing the metaverse-based model and the principles of the data economy in mind, we propose an innovative intersection of machines, robots, and Internet of Things (IoT) devices. These tools capture real-world data to construct multiple metaverses, or interactive platforms. These platforms enable users to embed new logics and Artificial Intelligence (AI) algorithms, contributing to the design process through blockchain-based consensus protocols.

In these virtual worlds, AI is guided by human-defined logics, learning from interactions and assessing designs through consensus, thereby establishing a reward system for platform participants. The resulting intelligence can command machines in the real world to perform tasks that range from hazardous operations during critical situations to assisting humans in difficult tasks and automating industries.

Projecting this model to a disaster scenario, we imagine simulating an earthquake in a virtual town and illustrate how the cycle of interaction, learning, and improvement can expedite and fortify real-world disaster response and reconstruction. We also discuss the model's critical aspects, including potential security threats, data integrity issues, technological limitations, ethical AI considerations, and issues of accessibility and participation.

Embedding the cultural and theoretical context of late 20th-century architectural thought, this section contributes to the conversation on the role of AI and metaverses in urban reconstruction and disaster management. By envisioning future scenarios for emergency situation management, we underscore the practical potential of these concepts and their implications for resilient urban development.

5.2 *Theoretical Background: Intersecting Lebbeus Woods' Concepts with Artificial Intelligence in Contemporary Architecture*

This study is grounded in the architectural theories of Lebbeus Woods, specifically his concepts of *heterarchy* and *cyclicity* [22]. Woods' work challenges traditional hierarchies and linear models of development, advocating instead for a more complex and dynamic understanding of architectural and urban forms. His *Reconstruction Series* [23] provides a conceptual framework for this study, offering a radical rethinking of how we approach the built environment, particularly in the aftermath of disasters.

Heterarchy, a key concept in Woods' work, refers to the organization of elements where no one element has an inherent superiority over the others. This concept challenges the hierarchical structures often found in traditional architectural and urban planning practices, both top-down and bottom up. *Cyclicity*, on the other hand, refers to a non-linear understanding of time and development, where events and processes are seen as part of an ongoing cycle of destruction and reconstruction. This concept offers a way to think about the continuous evolution and transformation of our built environments.

5.3 *Hetherarchy and Cyclicity: Relationships with AI*

The concepts of heterarchy and cyclicity can indeed find interesting parallels and applications in the realm of artificial intelligence (AI). This discussion extends not only to AI architecture and operations but also to the ways AI may influence societal structures and our relationship with time.

Beginning with heterarchy, the non-hierarchical organization of elements where no one component has inherent superiority is particularly relevant in decentralized AI and machine learning models. A heterarchical approach enables the use of 'ensemble learning', where multiple independent models collaborate, each contributing their unique strengths, and compensating for the weaknesses of others. This *differentiation* leads to a more robust and accurate model, reflecting the way elements in a heterarchical design can work together to create a *stronger whole.*

Furthermore, as AI becomes increasingly embedded in our societies, the heterarchical model may challenge conventional power dynamics. Instead of a centralized, top-down control of AI, where few entities have dominance, a heterarchical arrangement could distribute AI capabilities more equitably, fostering a broader, more democratic access and engagement with this technology.

Cyclicity, on the other hand, finds a resonance in the iterative processes at the heart of AI and machine learning. Algorithms learn and evolve through repeated cycles of training, evaluation, and refinement. They are fed input data, make predictions, receive feedback about their accuracy, and then adjust their internal parameters accordingly, in a continuous process of destruction (of incorrect assumptions) and reconstruction (of refined understanding).

Moreover, the concept of cyclicity can apply to the broader societal implications of AI. AI is not a static development but rather a catalyst of ongoing cycles of *transformation*, *disruption*, and *adaptation*. AI itself learns and evolves in cycles, and its increasing integration into societies leads to cyclical transformations in work, leisure, education, healthcare, and virtually all aspects of life.

5.4 Bridging AI and Architecture: A Critical Perspective

Having delved into the intersections of AI with heterarchy and cyclicity, we can now appreciate how these dynamics apply to practical scenarios, notably post-disaster reconstruction. Woods critiques the prevailing reasoning system, often embodied in rigid planning and design practices, arguing that it can limit our ability to respond creatively and effectively to the intricate challenges presented by disasters. By moving away from the application of an abstract and universal system of form-making and space modeling, we can foster a more flexible, adaptive approach. Drawing from Woods' critiques and the theoretical concepts explored, we propose a novel solution: a metaverse-based model for post-disaster reconstruction. Details of this innovative approach will be discussed in the upcoming sections.

6 Methodology: A Metaverse-Based Model for Post-Disaster Reconstruction

6.1 Conceptual Framework

Our methodology for post-disaster reconstruction is rooted in the creation of a metaverse-based model, which is a virtual reality space where users can interact with a computer-generated environment and other users. This section outlines the key components of this methodology (see Fig. 1).

6.2 Data Acquisition from the Real World: Role of Machines, Robots, and IoT.

The first step in our methodology involves the acquisition of data from the real world. This is achieved through the use of machines, robots, and Internet of Things (IoT) devices, which are capable of capturing a wide range of data types and volumes. This data forms the basis for the construction of our virtual worlds or metaverses.

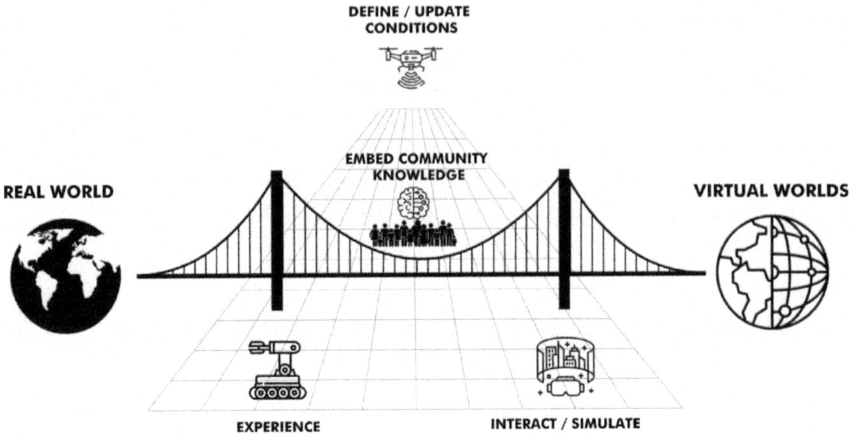

Fig. 1 A bridge between two worlds: reality and virtuality. The pillars symbolize two fundamental aspects—interaction in virtual worlds and remote control and automation of devices and machines in the real world. The catenary represents the involved communities, engaged in a continuous dialogue to generate and regulate the intelligence necessary for this process. Above, the drone represents the constant capture and updating of information, essential activities to create the ideal environment to first simulate and then implement future automation scenarios

6.3 Constructing Virtual Worlds (Metaverses): Algorithms and Models.

Following the acquisition of real-world data, this information serves as the building blocks for creating virtual worlds or metaverses. The process employs an hybridization of game engines, CAD software, and a combination of hard-coded and visual-based programming languages.

Game engines pave the way for the creation of vibrant 3D environments, while CAD software uses the real-world data to design accurate models. Hard-coded programming languages, traditionally the domain of computer scientists, represent one side of the programming spectrum.

However, the crucial role of visual-based languages, like those harnessed on platforms such as Grasshopper for Rhinoceros or Blueprint for Unreal Engine, cannot be understated. These languages stand out due to their intuitive, visual programming interfaces. They democratize the process of creating algorithms and writing code, extending their use to a broader user base. This includes creatives and professionals from non-technical backgrounds, who can easily process and interpret data to create a virtual world that faithfully mirrors reality. This fosters a more inclusive and diverse user community, playing a pivotal role in the expansion of the metaverse.

6.4 Human Interaction and Logic Embedding in the Virtual World.

Within these metaverses, users can interact with the virtual environment and with each other.

Their actions can also instruct new logics and Machine Learning (ML) algorithms, which can be used to simulate learn from various scenarios and outcomes. This allows for a high degree of flexibility and adaptability, as the metaverse can be continually updated and modified based on user interactions and inputs.

6.5 Emergence of a New Level of Intelligence: Theoretical and Practical Implications.

As users, designers and programmers interact with the metaverse and embed new logics and AI algorithms, a new level of intelligence begins to emerge, transforming the platform into a stochastic, event-based system.

This shift signifies a move from *emergency* management to *emergence* management, where the system's responses are not just pre-programmed and deterministic reactions, but are dynamically shaped by ongoing interactions and events. This *emergent intelligence*, a product of human–machine interaction, has the potential to revolutionize the way we approach post-disaster reconstruction. On a theoretical level, it challenges traditional notions of intelligence and cognition. On a practical level, it offers the possibility of more efficient, effective, and responsive disaster management strategies, adapting in real-time to the complexities of the situation. This metaverse-based model for post-disaster reconstruction represents a significant departure from traditional methodologies. It offers a more dynamic, interactive, and adaptable approach, which is better suited to the complex and unpredictable nature of disaster and more generally real life scenarios.

6.6 Case Study: Potential Application of the Model to a Disaster Scenario

The practical application of our metaverse-based model for post-disaster reconstruction can be illustrated through a potential case study involving a simulated disaster scenario.

6.7 Reality Capture and Virtual Model Creation: The Role of Drones

The first step in this potential case study could involve the use of drones to capture real-world data from a specific location.

Drones can capture raw data, which is then transformed into a 3D representation known as a point cloud. This detailed model captures the precise characteristics of a real-world location, like a town.

However, point clouds can be data-heavy, making them difficult for computer models to process efficiently. To alleviate this, *voxelization* [24] is employed. Voxelization simplifies these complex point clouds into voxels, 3D equivalents of pixels. This process retains much of the original detail while making the data more scalable for computational handling at different levels.

The resulting voxelized model is then versatile enough for simulating various disaster scenarios, assisting in testing and improving disaster management strategies.

6.8 Simulating an Earthquake Scenario: Method and Results

In our virtual town, we can mimic a scenario like an earthquake using advanced software systems that simulate the impact on buildings and infrastructures. Integrating the voxel-based model with the high-end destruction simulation software creates a robust platform for disaster scenario testing.

Each voxel in the model—derived from the initial point cloud data—represents a distinct part of the town's infrastructure, acting as a small puzzle piece in the larger model.

In an earthquake simulation, the software applies complex algorithms to these individual voxels. This approach translates the simulated seismic forces onto each section of the town, creating a detailed landscape of potential damage. The software considers voxel-specific properties, like material strength and structural design, ensuring realistic outcomes. Thus, the merging of voxel architecture and advanced simulation software delivers a powerful tool for intricate and comprehensive disaster scenario modeling.

6.9 Player Interaction and Town Reconstruction: A Participatory Approach.

Once the earthquake simulation has been run, the damaged virtual city can be transformed into an interactive platform for players. This transition would see participants taking on the challenge of rebuilding the town. However, to make the task more engaging and to foster a sense of progress, specific goals, objectives, and scores can be set.

Players would be tasked with the efficient reconstruction of the town's infrastructure, with success measured through defined objectives. These could range from rehabilitating certain buildings within a set timeframe, to reestablishing vital services like power or water supply, to ensuring the overall livability of the town. Progress towards these goals can be quantified through a scoring system that awards points based on effectiveness, speed, sustainability, and creative problem-solving in the reconstruction efforts.

This gamified approach not only encourages creativity and teamwork among the players, but it also allows for the exploration of various disaster recovery strategies in a dynamic, interactive setting. It adds an educational element to the task, creating a powerful tool that bridges the gap between disaster management theory and practice.

6.10 Continuous Improvement Through Repeated Cycles: A Cyclical Model of Learning.

As players interact with the virtual environment and participate in the reconstruction process, the system could continually learn from their actions. This cyclical model of learning would allow for continuous improvement, as the system adapts and evolves based on the players' inputs and feedback.

6.11 AI Learning from User Logics: Method and Implications.

The logics defined by the players in the virtual world could instruct the AI, enabling it to learn from the players' interactions and decisions. This process could have significant implications for the development of more effective and efficient disaster management strategies.

In this context, blockchain technology could play a pivotal role in incentivizing and rewarding players who provide more meaningful information to the platform. Blockchain, a decentralized and transparent ledger system, could be used to create a token-based reward system within the metaverse. As players interact with the virtual environment and contribute their unique logics and strategies, these contributions could be tracked and verified on the blockchain. Each meaningful contribution could be rewarded with tokens, which could be traded within the platform or potentially exchanged for real-world value. This blockchain-based reward system could encourage players to engage more deeply with the platform and to contribute their best ideas and strategies. It would also ensure transparency and fairness, as the blockchain would provide a verifiable record of each player's contributions.

Moreover, the use of blockchain could also facilitate the creation of decentralized autonomous organizations (DAO) involved in specific tasks within the metaverse. These DAOs could be governed by the players themselves, with decision-making powers distributed according to the number of tokens each player holds. This could further incentivize meaningful participation, as players would have a direct stake in the governance and success of the platform. In this way, blockchain technology could enhance the effectiveness of the metaverse-based model for post-disaster reconstruction, by incentivizing meaningful participation, rewarding valuable contributions, and fostering a sense of ownership and community among the players.

6.12 Real World Disaster Response and Reconstruction: A Practical Application.

The final stage of our potential case study could involve the application of the lessons learned from the virtual world to a real-world disaster scenario. The strategies and solutions developed in the virtual world could inform the response and reconstruction efforts in the real world, demonstrating the practical value of our metaverse-based model.

In this context, the information collected in the metaverse could be used to create the code to instruct autonomous robots and vehicles to work autonomously during emergency phases. This would represent a significant shift from *improvisation strategies* to *predictive emergence strategies.*

As players interact with the virtual environment and respond to simulated disaster scenarios, their actions and decisions could be recorded and analyzed. This data could then be used to train AI models, which could learn to predict and respond to similar situations in the real world.

For example, if players consistently use a particular strategy to evacuate people from a building during a simulated earthquake, the AI could learn to implement this strategy in a real-world earthquake scenario. This could involve instructing autonomous vehicles to follow a specific evacuation route, or deploying autonomous robots to carry out search and rescue operations.

Moreover, the AI could also learn from the mistakes and failures that occur in the virtual world. If a particular strategy proves ineffective or leads to negative outcomes in the simulation, the AI could learn to avoid this strategy in the real world.

In this way, the metaverse could serve as a virtual testing ground for disaster response strategies. The lessons learned in the virtual world could be used to inform and improve the actions of autonomous systems in the real world, leading to more effective and efficient responses to real-world disasters.

This approach could also help to reduce the reliance on improvisation during emergency phases. Instead of reacting to disasters in an ad-hoc and potentially chaotic manner, responders could draw on the lessons learned from countless simulated scenarios. This could lead to more coordinated, effective, and predictable responses, ultimately improving the safety and resilience of our communities.

6.13 Discussion: Critical Points of the Model

While the metaverse-based model for post-disaster reconstruction presents numerous opportunities for innovation and improvement, it is also important to consider the potential challenges and limitations of this approach.

6.14 The Threat of Hacking: Security Concerns

As with any digital platform, the metaverse is susceptible to the threat of hacking. Unauthorized access to the system could lead to the disruption of operations, manipulation of data, or even the creation of false scenarios. It is crucial to implement robust security measures to protect the system and its users.

6.15 Malicious Data Injection: Risks and Mitigation

Another potential risk is the injection of malicious data into the system. This could distort the virtual environment and the AI's learning process, leading to inaccurate or harmful outcomes. Strategies for detecting and mitigating malicious data injection need to be investigated, developed and incorporated into the system.

6.16 Technological Challenges in Real-Time Processing

The efficiency of a metaverse-based model relies heavily on real-time data capture and processing. However, substantial computational and network limitations can present obstacles. The high volume of data necessary for detailed, dynamic models pushes the limits of current hardware and requires sophisticated, optimized algorithms. Simultaneously, data transmission needs high-speed internet infrastructure; any latency or bandwidth issues can compromise user experience. Therefore, to achieve a seamless metaverse experience, there is an essential need for more powerful hardware, refined software, and robust networking.

6.17 *Ethical AI Issues*

The use of AI in the metaverse raises a number of ethical issues. These include concerns about privacy, consent, accountability, and the potential for bias in the AI's learning and decision-making processes. A critical examination of these issues is necessary to ensure that the system is not only effective but also ethical.

6.18 *Exclusion of Some People: Issues of Accessibility and Participation*

Finally, there is the risk of excluding some people from the metaverse, particularly those who lack the necessary technology or skills to participate. This could limit the diversity and inclusivity of the system, and potentially reinforce existing inequalities. Strategies for promoting accessibility and participation need to be an integral part of the system's design and implementation.

These critical points highlight the need for careful planning, ongoing evaluation, and continuous improvement in the development and implementation of every metaverse-based model, and specifically for post-disaster reconstruction.

7 Conclusion

The exploration of a metaverse-based model for post-disaster reconstruction presents a compelling approach to enhance community preparedness while managing and responding to disasters. It builds upon the understanding of cities as complex systems and the need for resilience in the face of environmental challenges. The potential of the cybernetically inspired metaverse-based model, influenced by visionary thinkers who have shaped cultural currents, harnesses the transformative power of AI, challenges and redefines traditional methodologies in architecture and design, allowing for dynamic, adaptive, and interactive environments that respond intelligently to changing conditions.

The potential application of this model, as illustrated through a hypothetical case study, demonstrates its capacity to revolutionize the field of post-disaster reconstruction. By creating an immersive, interactive virtual environment, we can simulate disaster scenarios, engage users in the reconstruction process, and learn from their interactions to continually improve our strategies.

However, the implementation of this model is not without challenges. Security concerns, the risk of malicious data injection, technological limitations, ethical AI issues, and accessibility are all critical points that need to be carefully considered and addressed.

Despite these challenges, the potential benefits of this metaverse-based model in bridging the gap between the virtual and the physical world are significant. By transforming the way we approach this integration, we can enhance our ability to respond to changes, improve the resilience of our digital and physical environments, and ultimately, create safer, more sustainable communities.

The metaverse-based model offers a comprehensive and interconnected virtual space that has the potential to reflect, and possibly augment, our physical reality. With this, it is conceivable that this model can be used to simulate and predict real-world scenarios, provide a platform for education and training, and even plan or test urban development projects before they are physically implemented.

As we progress, it will be crucial to continue exploring and refining this model, conducting further research, and engaging in ongoing dialogue with all stakeholders. This includes the technology developers, urban planners, policy makers, and the general public who would inhabit both these physical and virtual spaces. It's through these efforts that we can fully realize the potential of this innovative approach in bridging the gap between virtual and real to enhance urban resilience.

8 Biographical Notes

Graduated in Building Engineering Architecture, Carlo Beltracchi has dedicated his career to the fusion of architectural engineering and emerging technologies. Post-graduation, he pursued a path that integrates architectural engineering research with practical implementations, focusing on computational and material systems to propel mass-customization in the AEC industry via robotics and automation. Lately, his exploration has led him to the forefront of digital architecture, devoting his research to realms like Metaverse Architecture and Web3 applications. His efforts aim to establish a connection between traditional architectural engineering practices and the ever-expanding digital sphere, making significant contributions to research in this dynamic field, all while keeping a project-oriented approach.

References

1. Anderson PW (1972) More is different. Science (New York, N.Y.) 177(4047):393–396
2. Peterson G, Allen CR, Holling CS (1998) Original articles: ecological resilience, biodiversity, and scale. Ecosystems (New York, N.Y.) 1(1):6–18
3. DeLanda M (2005) Space: extensive and intensive, actual and virtual. In Deleuze and Space (pp. 80–87). Edinburgh University Press, Edinburgh
4. Cengel YA, Boles MA (2019) Thermodynamics: an engineering approach (9th ed.). McGraw-Hill Education, New York City
5. Holling CS (1973) Resilience and stability of ecological systems. Annu Rev Ecol Syst 4(1):1–23
6. World Economic Forum (2023) Global risks report 2023, pp 70–71
7. UN. Population Division (2018) The World's cities in 2018 : data booklet. UN

8. Grimm NB, Faeth SH, Golubiewski NE, Redman CL, Wu J, Bai X, Briggs JM (2008) Global change and the ecology of cities. Science (New York, N.Y.) 319(5864):756–760
9. Walker B, Holling CS, Carpenter SR, Kinzig AP (2004) Resilience, adaptability and transformability in social-ecological systems. Ecol Soc J Integ Sci Resil Sustain 9(2)
10. Zevi M (1972) Architettura & computer: Il computer nella progettazione. Bulzoni, Roma
11. Mohammadi M, Al-Fuqaha A (2018) Enabling cognitive smart cities us-ing big data and machine learning: approaches and challenges. IEEE Commun Mag 56(2):94–101
12. Banerjee A, Nayaka RR (2022) A comprehensive overview on BIM-integrated cyber physical system architectures and practices in the architecture, engineering and construction industry. Constr Innov 22(4):727–748
13. Nwaogu JM, Yang Y, Chan APC, Chi H-L (2023) Application of drones in the architecture, engineering, and construction (AEC) industry. Autom Constr 150(104827):104827
14. Buyuksalih I, Bayburt S, Buyuksalih G, Baskaraca AP, Karim H, Rahman AA (2017) 3d modelling and visualization based on the unity game engine—advantages and challenges. ISPRS Annal Photog Remote Sens Spat Inform Sci IV-4/W4:161–166
15. Lin Z, Yao N, Wu X, Wang L (2022) A peek at metaverse society from web 3.0 games: a preliminary case study of dark forest. 2022 IEEE 24th International Workshop on Multimedia Signal Processing (MMSP), pp 1–6
16. Bucher FD, Hall MD (2022) New ways of data governance for construction? Decentralized data marketplaces as Web3 concept just around the corner. Proceedings of the 29th EG-ICE International Workshop on Intelligent Computing in Engineering
17. Dounas T, Lombardi D, Jabi W (2021) Framework for decentralised architectural design BIM and blockchain integration. Int J Archit Comput 19(2):157–173
18. Cao L (2022) Decentralized AI: Edge intelligence and smart blockchain, metaverse, Web3, and DeSci. IEEE Intell Syst 37(3):6–19
19. Global B [@BomaGlobalStudio] (2019) Nature 2.0: a positive vision of the future | Trent McConaghy | Boma Germany summit 2019. Youtube. https://www.youtube.com/watch?v=trwbHJZQl44, Last accessed 2023/07/17
20. Hunhevicz JJ, Wang H, Hess L, Hall D (2021) No1s1—a blockchain-based DAO prototype for autonomous space
21. Dounas T, Lombardi D, Vele J, Prokop S (2022) A crypto-twin framework for the AEC industry, Enabling Digital Twins with Blockchain technologies
22. Lebbeus Woods: the Vico Morcote interview (1998) [Video]. https://youtu.be/KO7-sGk6O8g, last accessed 2023/07/17
23. Woods L (1997) Radical reconstruction, 1997th edn. Princeton Architectural Press, NYC
24. Xu Y, Tong X, Stilla U (2021) Voxel-based representation of 3D point clouds: methods, applications, and its potential use in the construction industry. Autom Constr 126(103675)

Part II
Virtuous Workflows

Chapter 6
Towards a Rhizomatic Design Model

Michele Andaloro

Abstract Over the past three decades, as multiple layers of complexity, sustainability aspirations, innovative design and manufacturing methodologies emerged, the need for novel integrated design processes has become crucial. While the widespread introduction of computational design, building information modeling, and robotic fabrication has brought significant advances in the AEC industry, traditional linear models limited the potential of these tools. On the other hand, rethinking design models as a rhizome (ῥίζωμα, rhízōma, "mass of roots") can accomplish a non-hierarchical network that "connects any point to any other point" (Deleuze and Guattari), where digital tools can provide a powerful platform to carry out a truly interdisciplinary collaboration. Two aspects are particularly illustrative of this model: sustainability, where the organic embedment of sustainable measures in the design process yields more meaningful results and a higher-level functionality throughout an object's lifecycle, and fabrication, where the integration of digital and robotic technologies generates efficiencies in terms of production and data management, but it also fosters innovative material explorations and architectural expressions.

Keywords Rhizomatic design model · Non-hierarchical network · Sustainability · Lightweight · Lifecycle

1 Introduction

The rising complexity of the building process, encompassing all disciplines and stages, and the fact that technical matters are increasingly important—due to high performance and sustainability standards as well as the sophistication of the architectural systems—require an elevated degree of specialization of the agents involved. This has brought the need for novel collaborative design models to ensure that the process does not lead to an assemblage of ultra-complex vertical silos, lacking in

M. Andaloro (✉)
Werner Sobek, New York, USA

P. Ruttico (ed.), *Coding Architecture*, Digital Innovations in Architecture, Engineering and Construction, https://doi.org/10.1007/978-3-031-47913-7_6

interrelation, but rather to an integrated and multidisciplinary system for the definition of a project.

Digital protocols supported this advancement in the industry: they allowed projects of different scales and programs to deal with an increasing level of complexity and meet higher architectural and performance standards; however, they have been primarily implemented following traditional linear models—multiform, but inherently hierarchical—thus not allowing, due to their internal structure, a complete expression of their potential.

The body of data that is generated and the possibility to share these inputs throughout the whole process and across disciplines are still limited to specific instances and the structure of the overall model—constituted of defined steps and compartmentalized—limits the effectiveness of integrated feedback and mutual exchanges.

This chapter provides a novel reading of the digital advances in the AEC industry, aimed at fostering a broader interconnection between tools, a more integral reciprocity between disciplines, and a move beyond a traditional linear workflow towards a horizontally integrated and distributed design model.

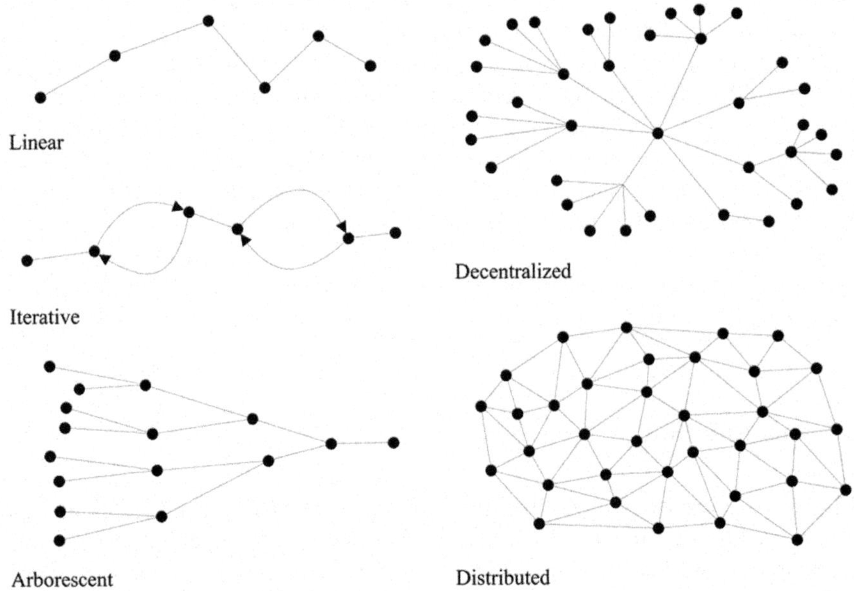

Fig. 1 Examples of propagation models

2 Beyond Linearity

To achieve a higher level of integration, there is a need to reconsider conventional systems to effectively deliver architectural projects and establish a network of interconnected dependencies.

This model can be imagined as a rhizome (ῥίζωμα, rhízōma, "mass of roots"): a non-hierarchical network that "connects any point to any other point" (Deleuze and Guattari) and—going beyond traditionally compartmentalized stages and disciplines—allows an entangled design process where the result emerges from informed interactions rather than from a centralized source.

The nodes of the rhizome—disciplines, phases, stakeholders, aspirations, and performances—create a network of systemic dependencies and interactions, a generative model based on growth and propagation. The horizontal concatenation is potentially infinite, expanding beyond traditionally separated phases and achieving a fully integrated system.

Emerging creative endeavors in the last century are demonstrating how invention and innovation reside in the interrelation between traditionally separated disciplines. In this light, the systemic and widespread application of a distributed model in the building sector—inherently multidisciplinary and involving a multitude of aspirations—can provide the means to develop novel solutions to shape the built environment and achieve meaningful results, especially from an environmental perspective.

The resulting concatenation is extensive, and the lack of a clear hierarchical structure makes it look inextricable; however, the potential to create a truly creatively

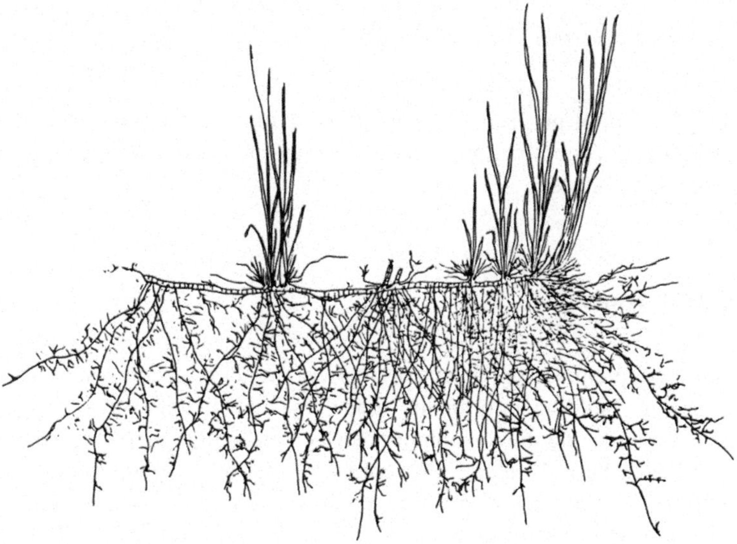

Fig. 2 Illustration of a rhizome

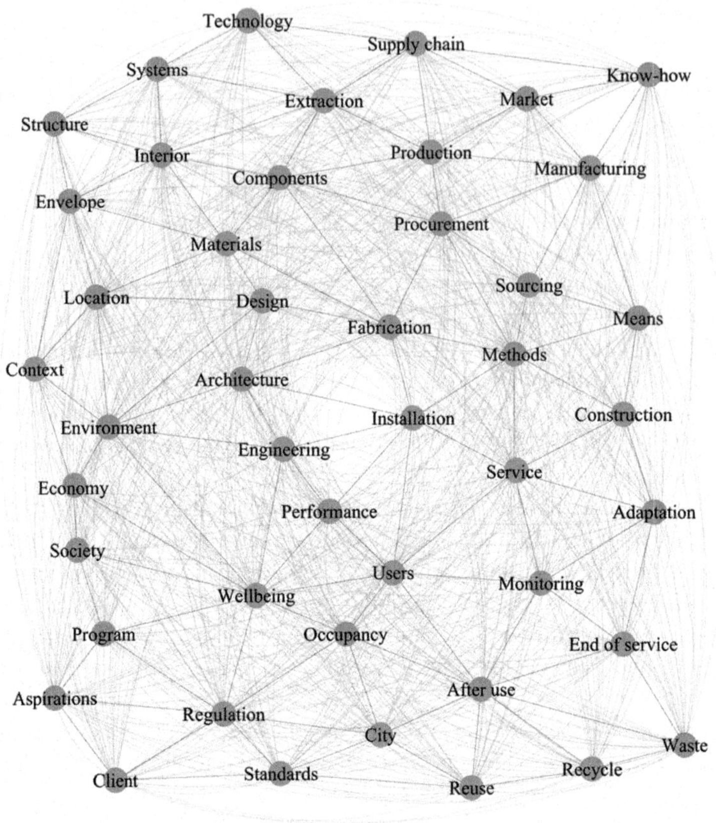

Fig. 3 Schematic diagram of a rhizomatic model

flourishing system of propagation in architecture urges the development of a technical mapping to encompass this multiplicity and to enable designers to operate project-specific syntheses.

3 Digital Tools

Under different premises and objectives, current digital protocols—such as computational design, building information modeling, and digital twin technologies—fostered integration and interoperability between disciplines and widened the spectrum of influence between phases. Nevertheless, a further interconnection is needed to accomplish a full horizontal integration between all nodes and ensure that data are effectively propagated and can inform all the agents in the process.

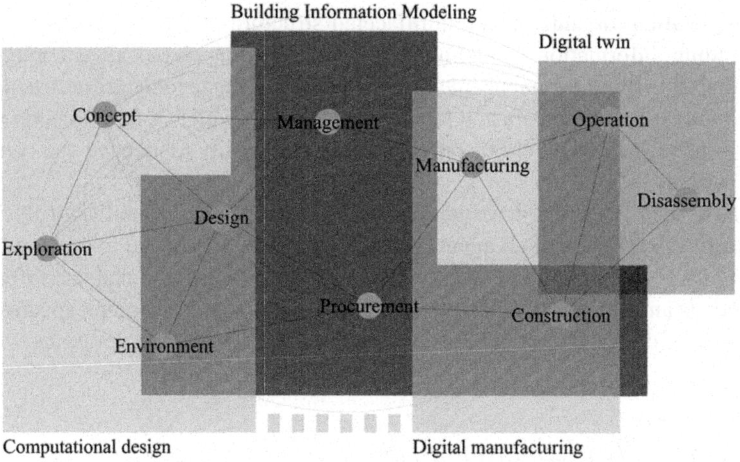

Fig. 4 Potential integration between contemporary digital tools

3.1 Computational Design Tools

Computational design set out a fundamental shift in design and engineering, replacing representation, Computer Aided Design, with generation and simulation; this allowed designers to integrate increasingly sophisticated sets of information across all disciplines and progressively reduce the distance between design and fabrication.

From a design perspective, computational tools—whether parametric, generative, or algorithmic—augmented the design possibilities in architecture, both technical and compositional, and enabled a deeper exploration of geometrical forms, materials, and environmental agents, making it possible for all these data to be embedded and actively inform the result via designer-generated codes.

On the other hand—in the context of a widespread introduction of numerically controlled, digital, and robotic fabrication methodologies—computation provided a linguistic and logic connection between conceptualization and materialization. Indeed, it allowed an active cross-pollination of information between these stages: designs are informed by the tools and the methodologies that will be implemented in their realization, while the design inputs can be turned into fabrication instructions and means to organize the production.

3.2 Building Information Modeling

Building Information Modeling tools enabled the systemic embedment of information in design models, creating an augmented digital representation of projects that efficiently carries and manages a complex and multidisciplinary system of data through the different stages.

The resulting organized and multifaceted structure of the data fostered the use of Building Information Modeling beyond the design phases to encompass the entire building lifecycle management—including cost management, construction and project management, system and facility operation, as well as environmental assessments—and provided all entities in the process the ability to manage and centralize information.

Whilst such platforms have undoubtedly enhanced the capability of designers, architects, engineers, and all agents to collaborate efficiently and reliably, the interoperability of these platforms with computational protocols and analysis software—at all levels and between disciplines—is still limited from a design perspective, thus not allowing them to be a powerful instrument for generative explorations.

3.3 Digital Twin and Adaptivity

The extension of the scope of digital modeling beyond the design and construction phases and the integration of real-time usage and operational data led to the introduction of digital twins in the AEC industry. As a synchronized digital counterpart of real objects, they can be used as a bi-directional connection between the digital and the physical world for monitoring, simulation, integration, testing, prediction, and operation purposes to achieve an increasingly accurate product lifecycle management based on analytical data.

The synchronization of the building with its digital doppelganger allows the integration of real-time data within a holistic learning model that analyzes actual real-time information and refines operational setups.

This active interconnection is crucial for the effective widespread application of adaptive systems and overall to make buildings capable of optimizing their behavior according to point-in-time data, changing environmental conditions, and user requirements.

3.4 Future Development

The creation of a digital infrastructure to support a full horizontal concatenation—from the conception of a project to its disassembly—can open novel possibilities for the whole sector to build more efficiently, more sustainably, and more creatively.

As information is increasingly available and their propagation is traceable, two fundamental patterns can be identified: data-informed sustainable strategies moving backward to be organically embedded in the design concept and project data—materials, systems, context—moving forward and linking design and fabrication.

4 A Novel Approach to Sustainability

As sustainability is often relegated to mere environmental assessments or localized optimizations that have a limited impact on the overall design, the results are buildings where the environmental costs of design choices are not sufficiently taken into consideration or where the focus is solely on technical performance.

Starting from the fundamental understanding that not only does the building sector stand for approximately 40% of all emissions worldwide—about 12 billion tons of CO_2 every year—but also approximately 60% of natural resource consumption and more than half of the worldwide waste generation, it is essential to focus on the design of better-performing buildings, but also to take into proper account how resources are used throughout the whole process.

In this scenario, the ability to design in a network of systemic dependencies helps to think holistically about sustainability and integrate meaningful resource planning in the earlier stages of the design to inform the conceptualization of a project.

Furthermore, as sustainability data are more structured and increasingly accessible, the focus on resources—embodied emissions from construction are responsible for around 20% of global warming—can help push the boundaries of conventional architectural and engineering solutions and provide a viable path to lighten the footprint of the built environment.

4.1 Building Lightweight

While lightweight design has heavily influenced the design language and the technical advancement of the last three decades from an engineering and architectural standpoint, what makes this approach particularly relevant today is that it is inherently sustainable. Indeed, the applications of this paradigm are manifolds and enable designers to tackle the problem of resources—embodied, operational, and waste—and their responsible use.

A horizontally integrated model allows to refine engineering analysis and simulations, thus supporting the achievement of an absolute minimization in the use of both materials and energy—carefully evaluating the balance between embodied and operational energy on a project-specific basis—through design ingenuity. The same can be applied to the material/component selection and sourcing, as an information-driven model supports more in-depth sustainability evaluations and inform design decisions.

Looking at the whole material lifecycle, a lightweight approach, based on design and engineering sophistication, can also be applied to the assembly of technological components and buildings overall; in fact, a strict logic of building-for-disassembly—material shall produce no waste and be completely reused or recycled—intertwines coherently with the concept of building lightweight: with less, more consciously.

4.2 Human-Centric Design

Integrating performance-related parameters more organically in the design process and fostering better interoperability between disciplines can also achieve higher standards of wellbeing as the relationship between a space and its users can be explored more extensively through entangled and multifaceted information.

Traditional models are often based solely on the compliance with performance standards; in contrast, a holistic approach based on horizontal integration can pursue a "more fundamental and universal" concept (Keeling et al.) of wellbeing, encompassing a more comprehensive range of aspects than the ones focusing only on indoor comfort parameters. An example of a broader and more systemic approach to this topic is the "nine categories of overall well-being" (Hanc et al.) which complement the aspects related to indoor environmental comfort with further descriptors of users' wellbeing in buildings; these include: environmental quality, satisfaction, comfort, mental and physical health, hedonic and eudaimonic well-being, social well-being, and productivity.

5 Design is Fabrication

Conventional procurement schemes do not entirely allow the design ingenuity to be transferred into the construction phase as there is a strict separation between the design and the fabrication stages; also, due to scope responsibility issues, the coordination is often limited and some design inputs engineering analyses, assembly, or design studies are either lost in the process or duplicated.

The convergence between design, analysis, and fabrication allows designers to conceive highly articulated processes with continuous flows of information between design and materialization. In fact, the reciprocity between design and realization through the use of digital platforms robotic and digital manufacturing technologies, computational design can achieve significant results in terms of production/data management mass customization as well as assembly and material efficiency due to the cross-pollination between analysis and manufacturing.

Furthermore, by extending the scope of the material research and rethinking standard supply chain models strategies for reusability, recyclability and waste minimization can be integrated at the forefront of the design, leading to a more careful use of the resources.

6 Conclusions

The deep interconnection of design, material, technical, fabrication, sustainability, and operational aspects create a complex system coexistence that requires designers to operate a synthesis.

While an integrated digital infrastructure provides the means to organize and navigate this network of interactions, designers can unfold novel possibilities to pair research with creativity, development with sustainability, and shape an architecture of entangled rhizomes.

References

1. Antonelli P, Oxman N (2020) Material ecology. The Museum of Modern Art, New York
2. Barazzetta G (2015) Digital takes command: Orizzonti di progettazione e produzione digitale. Rubbettino, Soveria Mannelli
3. Deleuze G, Guattari F (2004) A thousand plateaus. Continuum, London
4. Hanc M, McAndrew C, Ucci C (2019) Conceptual approaches to wellbeing in buildings: a scoping review. In: Building research & information, vol 47, issue 6. Taylor & Francis, Abingdon, pp 767–783
5. Heinlein F (2019) Recyclable by Werner Sobek. av edition, Stuttgart
6. Heinlein F, Sostmann M (2015) Werner Sobek: light works. av edition, Stuttgart
7. Heisel F, Hebel DE, Sobek W (2019) Resource-respectful construction: the case of the Urban Mining and Recycling unit (UMAR). In: IOP conference series earth and environmental science, vol 225. IOPscience, Bristol
8. IPCC (2022) Climate change 2022: mitigation of climate change. IPCC—Intergovernmental Panel on Climate Change, Geneva
9. IPCC (2022) Climate Change 2022: Impacts, Adaptation and Vulnerability. IPCC — Intergovernmental Panel on Climate Change, Geneva.
10. IPCC (2021) Climate change 2022: The physical science basis. IPCC—Intergovernmental Panel on Climate Change, Geneva.
11. Keeling T, Clements-Croome D, Luck R, Pointer R (2012) How the sensory experience of buildings can contribute to wellbeing and productivity. In: Proceedings of 7th Windsor conference. The changing context of comfort in an unpredictable world. Windsor Conference, Windsor
12. Kepes G (1965) Structure in art and in science. George Braziller, New York
13. Kwinter S (2007) Far from equilibrium: essays on technology and design culture. Actar, Barcelona
14. Leistner S, Honold C, Maierhofer M, Haase W, Blandini L, Sobek W, Roth D, Binz H, Menges A (2020) Research on integral design and planning processes for adaptive buildings. In: Architectural engineering and design management, vol 18, issue 3. Taylor & Francis, Abingdon, pp 241–260
15. Menges A (2012) Material computation: higher integration in morphogenetic design. Wiley Academy, London
16. Menges A, Ahlquist S (2011) Computational design thinking. Wiley Academy, London
17. Menges A (2015) Material synthesis: fusing the physical and the computational. Wiley Academy, London
18. Rawsthorn A, Antonelli P (2022) Design emergency: building a better future. Phaidon Press, New York
19. Schlegl F, Honold C, Leistner S, Albrecht S, Daniel Roth D, Haase W, Leistner P, Binz H, Sobek W (2019) Integration of LCA in the planning phases of adaptive buildings. In: Sustainability, vol 11. MDPI, Basel

20. Sobek W (2022) Non nobis—über das Bauen in der Zukunft Band 1: ausgehen muss man von dem, was ist. av edition, Stuttgart
21. Sobek W (2023) Non nobis—über das Bauen in der Zukunft Band 2: Die Randbedingungen des Zukünftigen. av edition, Stuttgart
22. Sobek W (2020) 17 theses. In: Zumtobel group annual report 2019/2020 of Zumtobel Group. Zumtobel Group, Dornbirn
23. Sobek W(2016) Ultraleichtbau/ultra-lightweight construction. In: GAM.12 architecture magazine. Birkhäuser, Basel, pp 156–167
24. Thompson D (1992) Crescita e forma. Bollati Boringhieri, Turin
25. UNEP (2020) 2020 global status report for buildings and construction: towards a zero-emissions, efficient and resilient buildings and construction sector. UNEP—United Nations Environment Programme, Nairobi
26. UNFCCC (2016) Paris agreement. UNFCCC—United Nations Framework Convention on Climate Change

Michele Andaloro is a Building Engineer from Politecnico di Milano, working as an Associate at Werner Sobek New York. After graduating in 2015 with a thesis on algorithmic design and robotic fabrication—developed at Indexlab in the framework of the "Digital Takes Command" exhibition by Triennale di Milano—he joined Werner Sobek in New York, where he focuses on design, engineering, and sustainability. He is responsible for managing and leading the design of complex projects internationally; his portfolio includes academic and cultural buildings for major institutions—such as Harvard, ITT, FIT, and the Smithsonian—art installations, and a wide range of public and private projects across different scales and programs. He has been a lecturer, guest critic, and consultant in numerous universities, including Politecnico di Milano, The Cooper Union, The University of Pennsylvania, Pratt Institute, and Rice University.

Chapter 7
Enhancing Sustainability: Concept Design Optimization Workflow for Human-Centered Low Carbon Solutions

Alessandro Baldini, Claudia Brisolin, and Emiliano Adam Cerutti

Abstract Daylight, glare, and overheating are crucial factors in achieving optimal user comfort in buildings. However, the lack of balance in these aspects often leads to visual and thermal discomfort. To address this challenge, a new workflow is proposed to support designers in developing solutions that provide ample natural light while mitigating glare and overheating through iterative parametric modelling. This approach enables efficient comparison of a large number of daylight and irradiance simulations using simplified shoebox models with varying geometries, opaque and transparent components, solar control strategies, and glass properties. The workflow automatically eliminates configurations with excessive shading based on the minimum depth of well-lit areas in the room, allowing designers to determine a range of optimal facade configurations that align with the architectural vision and assist mechanical engineers in their computations. A practical example demonstrates how the workflow was applied to achieve high daylight comfort and reduced solar gains in a subtropical climate for a 220 m high office building. Additionally, the study suggests steps for quantitatively assessing operational embodied carbon reduction before and after implementing this approach.

Keywords Natural daylight · Glare · Overheating · Passive low carbon facade design · Human centric design

A. Baldini (✉) · C. Brisolin · E. A. Cerutti
Eckersley O'Callaghan Ltd, 236 Gray's Inn Road, London WC1X 8HB, UK
e-mail: alessandro@eocengineers.com

C. Brisolin
e-mail: claudia@eocengineers.com

E. A. Cerutti
e-mail: emiliano@eocengineers.com

P. Ruttico (ed.), *Coding Architecture*, Digital Innovations in Architecture, Engineering and Construction, https://doi.org/10.1007/978-3-031-47913-7_7

1 Introduction

Modern buildings are nowadays designed according to very detailed thermal and visual comfort analysis simulations, however still several research papers have analyzed and shown how many built projects often experience users' discomfort due to either poor natural lighting [13], glare [7], or overheating phenomena [15]. Achieving the optimal balance between visual comfort and overheating mitigation within the facade design requires a well-organized coordination process between mechanical engineers, facade engineers and architects. This makes the whole facade design a complicated multi-level coordination process. On one level there is a "phenomenon coordination" which is assessed by mechanical and facade engineers who must find the optimal balance between daylight, glare and overheating by coordinating visual and thermal simulations. On another level, there is a "stakeholders' coordination" process of different people involved in the design phase who should design and deliver the most appealing and best performing building envelope.

Common practices normally assess daylight and glare at the same time with annual daylight simulations of the indoor environment; the challenge comes when overheating risk mitigation is added [12]. The main complexity of assessing visual and thermal comfort at the same time is that, although they are all "solar-driven" phenomena, there is no common thread between daylight and overheating. Daylight and glare are in fact luminous flux-related parameters with their own metrics to assess them; whereas, overheating is often the result of simultaneous high solar gains, high external air temperature and poor ventilation, which are usually monitored by mechanical engineers with their dynamic thermal modelling simulations.

Another degree of complexity of optimizing daylight, glare and overheating is that they have a contrapositive feature: solar control strategies normally prevent glare and overheating discomfort by reducing as much as possible the amount of direct sunlight and irradiance getting in the internal rooms. The consequence of this is that the deeper portions of the rooms tend to become dark and often need artificial lighting to compensate for the reduction of daylight availability [3]. For this reason, it is difficult to find the right balance and trend-off between these three phenomena without penalizing one excessively.

In this chapter, the authors propose a new workflow which compares several facade designs for highly transparent building in warm and hot climates in a holistic and time efficient way.

2 Background

2.1 *Visual and Thermal Comfort in Hot Climates*

The World Green Building Council estimated in 2019 that buildings nowadays account for 39% of energy-related carbon emissions globally—with operational carbon accounting for 2/3 of that and embodied carbon for the remaining [16]. In hot climates the main source of electricity consumption is represented by the cooling load, which accounts for approximately 65% of the total Energy Use Intensity for modern commercial high-rise buildings, with the remaining being typically attributed to lighting, electrical equipment, lifts/escalators, and others.

Highly efficient building envelopes are the key to reduce the whole-life carbon any project, with the integration in the facade of passive strategies (e.g., shadings, glass coatings or opening vents) that are able to reduce the cooling demand through natural ventilation or solar gains control, while stile supplying an adequate amount of natural daylight for the users' wellbeing.

Table 1 sums up the boundary conditions and the parameters mainly affecting visual and thermal comfort that the designer needs to consider for glare and overheating risk mitigation and natural lighting supplying (Figs. 1, 2).

Table 1 Boundary conditions and controllable parameters affecting visual comfort and overheating

Daylight, glare, and overheating boundary conditions (non-controllable) • Location • Climatic conditions • Surrounding buildings • Hour, day, and month considered • Position in the room	Daylight, glare, and overheating (controllable) parameters: • Building geometry • Internal layout • Glass components thermal/optical properties • Opaque components properties • Shading properties and solar control strategies

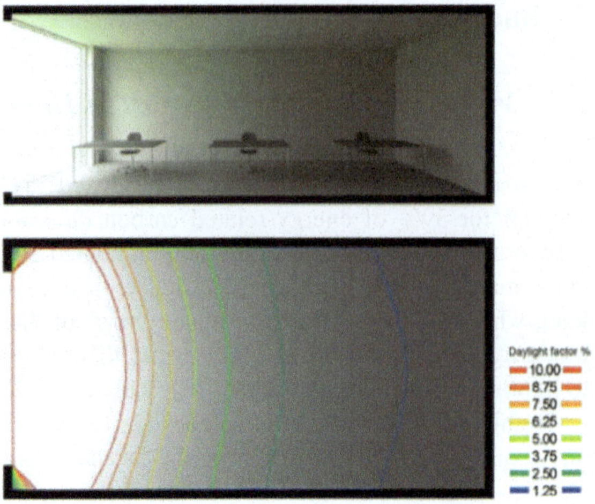

Fig. 1 30% of glazing to floor area ratio achieve a DF facade of 2% approximately 4.5 m from the facade. The DF average is equal to 5.1%, but it is highly non-uniform and not well distributed over the work plane area; this is a luminous environment likely to cause visual discomfort and glare. In this scenario, two of the three workplaces represented can be considered daylit [14]

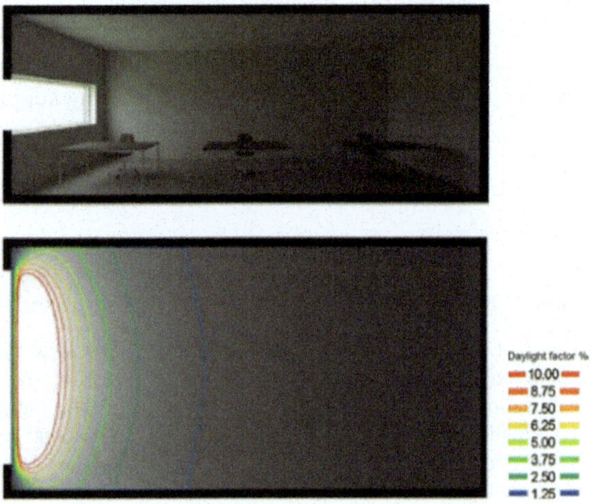

Fig. 2 10% of glazing to floor area ratio only achieve a DF of 2% a few metres from the facade and leave the back of the room with very low light levels. Even though the average DF of the room is equal to 1.9%, only a small work plane area achieves values above 2%, and only one of the three workplaces represented can be considered daylit. The other two require artificial lighting

2.2 Measuring Visual Comfort and Overheating Risk

Visual Comfort

Daylight comfort assessments often rely on Climate-Based Daylight Modelling (CBDM) simulations able to predict any luminous quantity (illuminance and/or luminance) at specific points of any internal space, at a specific time of the year. Several Illuminance-based metrics firstly reported by the Illuminating Engineering Society (IES) are currently used by designers: Uniformity Ratio (UR), Daylight Factor (DF), Daylight Autonomy (DA), Continuous Daylight Autonomy (CDA), Useful Daylight Illuminance (UDI) and Spatial Daylight Illuminance (sDA) [5].

Glare discomfort is usually assessed by measuring the Daylight Glare Probability in a specific point at a specific time of the year or by measuring the Annual Sunlight Exposure (ASE).

Several standards and protocols refer to the above-mentioned illuminance-based metrics and report different targets. BS EN 17037-2018+A1-2021 outlines the daylight simulations settings to be commonly used and reports the main requirements and targets to be achieved for visual comfort and glare mitigation in buildings. It sets for a number of CEN capital cities different Daylight Factor targets (D_T and D_{TM}) to be exceeded for more than half of daylight hours, over 50% of the reference plane.

BREEAM HEA 01 sets similar targets to BS EN 17037-2018+A1-2021 with target Daylight Factors (and illuminance over 50% of daylight hours) and minimum Target Daylight factors (and illuminance over 50% of daylight hours) over an area percentage between 60% and 80% in function of the latitude of the location and of the building type [2].

LEED v4.1 requires an average sDA300/50% greater than 40% for the regularly occupied area as a minimum daylight requirement and a maximum $ASE_{1000/250}$ of 10% for glare mitigation [11]. WELL v2, Q1-Q2 2023 reports a more stringent target of average $sDA_{300,50\%}$ of at least 55% of the regularly occupied floor area as a minimum requirement [10].

Overheating Risk Mitigation

Typically, the impact of the facade design onto the building's overheating risk can either be mitigated by adopting a prescriptive approach or it can be assessed through a detailed dynamic thermal modelling.

Prescriptive approaches normally rely on rules of thumbs which limit unwanted solar gains during the cooling season: Approved document O [9] for UK-based projects sets the maximum area of glazing in function of the floor area, of the elevation, of the risk location class and on the type of ventilation.

CIBSE's TM 52 [4], instead, relies on dynamic thermal modelling, and it sets a 3%-of-the-occupied-hours limit during which the operative temperature can exceed the threshold comfort temperature by 1K or more.

The novel approach introduced within this chapter relies on the assumption that a specific facade design significantly affects the perimeter zone only, with the interaction between facade and visual and thermal comfort becoming less evident in the deeper portions of any internal space.

Baker and Steemers first introduced the idea of passive zones in buildings and identified the first 6 m from the facade plane as the area which could potentially benefit from natural lighting, natural ventilation, and passive solar gains [1]. BSRIA Rules of thumbs for building services publication in 2011 set a 50–65 W/m2 threshold to the solar gains on a 4.5m deep perimeter zone [8].

3 Methodology

The workflow presented relies on a script that was developed to create shoebox-models which are located in strategic points of the facade, based on a preliminary solar exposure "zoning" of the building (typically at the most extreme high and low incident solar radiation hitting each elevation). While the geometry is simplified, shoebox-models can nonetheless consider local climate, facade performance parameters, daylight availability, solar loads, which makes them useful during schematic design to quickly study the effects of varying facade designs.

The shoeboxes are iteratively tested with different shading typologies and glass properties (geometry definition step) to measure the daylight depth, the glare depth and the irradiance hitting the floorplate perimeter zone (simulation step). An automatized process displays all the iterations on a triangular radar chart and selects the one achieving the best scores for the three parameters (post-processing step).

The computational stage needs to analyse different shoebox-models with short run times: CPU parallelization with Pollination is ideal to run simultaneously several daylight and energy simulations (this chapter reports 111 simulations which had been run in less than 3 h). Furthermore, the script records all the simulations based on their names and allows the user both to easily keep track of them and to share the simulation models—JSON files, which include geometries and model settings–with other engineers involved in the building design process to facilitate the coordination process of different teams.

The data-post processing stage carries out an automatic multi-criteria optimization process of different configurations (geometries, materials and building typologies) to identify the optimal facade design: the script supporting the workflow records all the results and finds the portion of room receiving an adequate amount of light (to be maximised) and the over-exposed (to be minimised) where the users may experience glare discomfort, while the incident irradiance on the floor plate is being reduced to avoid the risk of overheating and to reduce the cooling needs.

3.1 Geometry Definition

The geometry definition stage consists in creating a pool of different shoebox-models which vary both in their facade geometries and in their glass optical and thermal properties according to the input variables:

- Geometrical properties of the glazed components (window-to-wall ratio and windows locations)
- Optical and thermal properties of the glass components (glass g-value and light transmittance)
- External shadings geometrical properties (dimensions, number of shadings, inclination, arrangement)

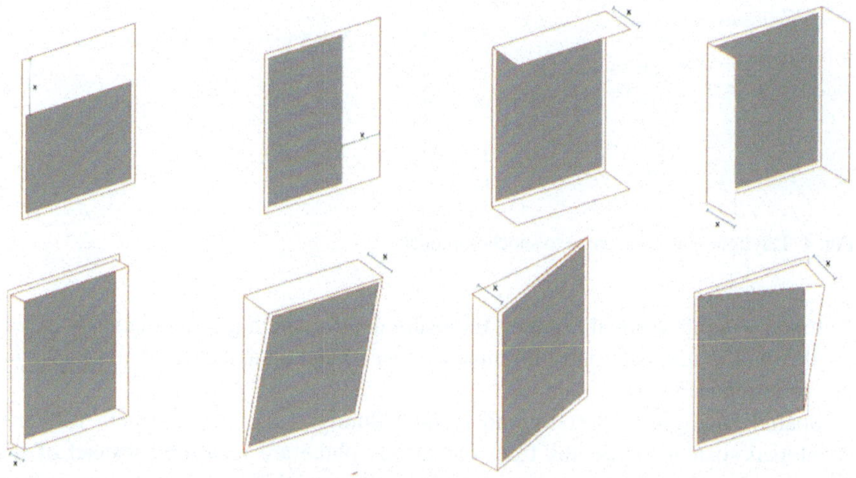

Fig. 3 Examples of parametric design for shadings

The workflow adapts to non-conventional shadings that can be parametrically implemented by the designer and included in the script (Fig. 3).

3.2 Simulation

The simulation stage consists in defining the boundary conditions and the settings for the solar radiation and daylight simulations. The shoebox-models are modelled with two different test meshes to assess both the visual and thermal comfort conditions in the perimeter zone: a 4.5m deep floor plate mesh (distance of 10cm from the floor level and offset from the walls equal to the grid size–typically 25cm for conventional shoebox-models) is used to measure the incident solar irradiance to assess the solar gains contributions, whereas a full floor-plate mesh (distance of 85cm from the floor level) is used to calculate the Spatial Daylight Autonomy depth and Annual Sunlight Exposure depth which are then post-processed by the algorithm to make them comparable among different facade designs.

3.3 Data Post-Processing

The data post-processing stage consists in a multi-criteria optimization process of the data collected from the different iterations defined in the first step and simulated in the second one: each model is associated with three values results–Spatial Daylight

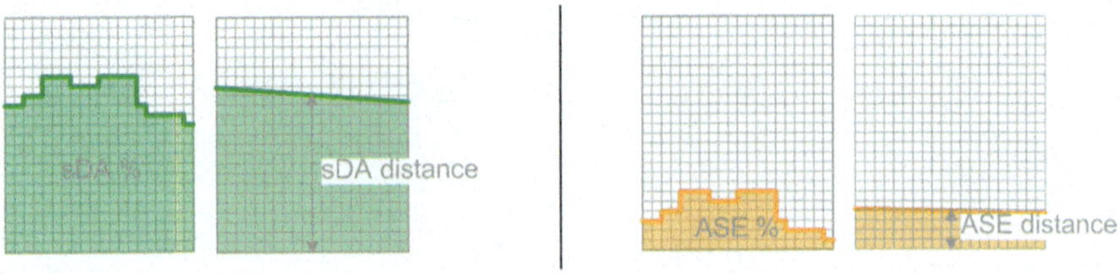

Fig. 4 Daylight metrics adapted to shoebox-models

Autonomy depth, Annual Sunlight Exposure depth (which are automatically calculated from the internal hourly illuminance levels) and annual average irradiance on the floor plate (Fig. 4).

Spatial Daylight Autonomy and Annual Sunlight exposure are both furtherly developed so that the depths from the facade plane are measured instead of the commonly used percentage value; this allows the workflow to not being constrained to the shoebox dimensions.

The three values are then normalized on a 0% to 100% scale according to the following arbitrary performance indexes which are used to create a radar triangular graph for each iteration.

$$A_i(daylight\ index) = \frac{(sDA\ depth)_i}{(ASE\ depth)_{MAX}}$$

$$B_i(glare\ index) = \frac{(ASE\ depth)_{MIN}}{(ASE\ depth)_i}$$

$$C_i(irr.index) = \frac{(IRR_{yearly\ mean})_{MIN}}{\left(IRR_{yearly\ mean}\right)_i}$$

The target is to maximise the three indexes by applying the previous formulas and comparing different facade options. The script automatically calculates the triangle's area for each iteration and identifies the optimal facade design with the one achieving the greatest area value (Fig. 5).

4 Application and Analysis

The novel workflow has been then applied on an exemplar high-rise building located in Hong Kong to test it in hot weather conditions and in a high-density urban area. The aim is to replicate the typical design process of a highly transparent skyscraper and to simulate the effect of shadings and obstructions on the internal comfort conditions.

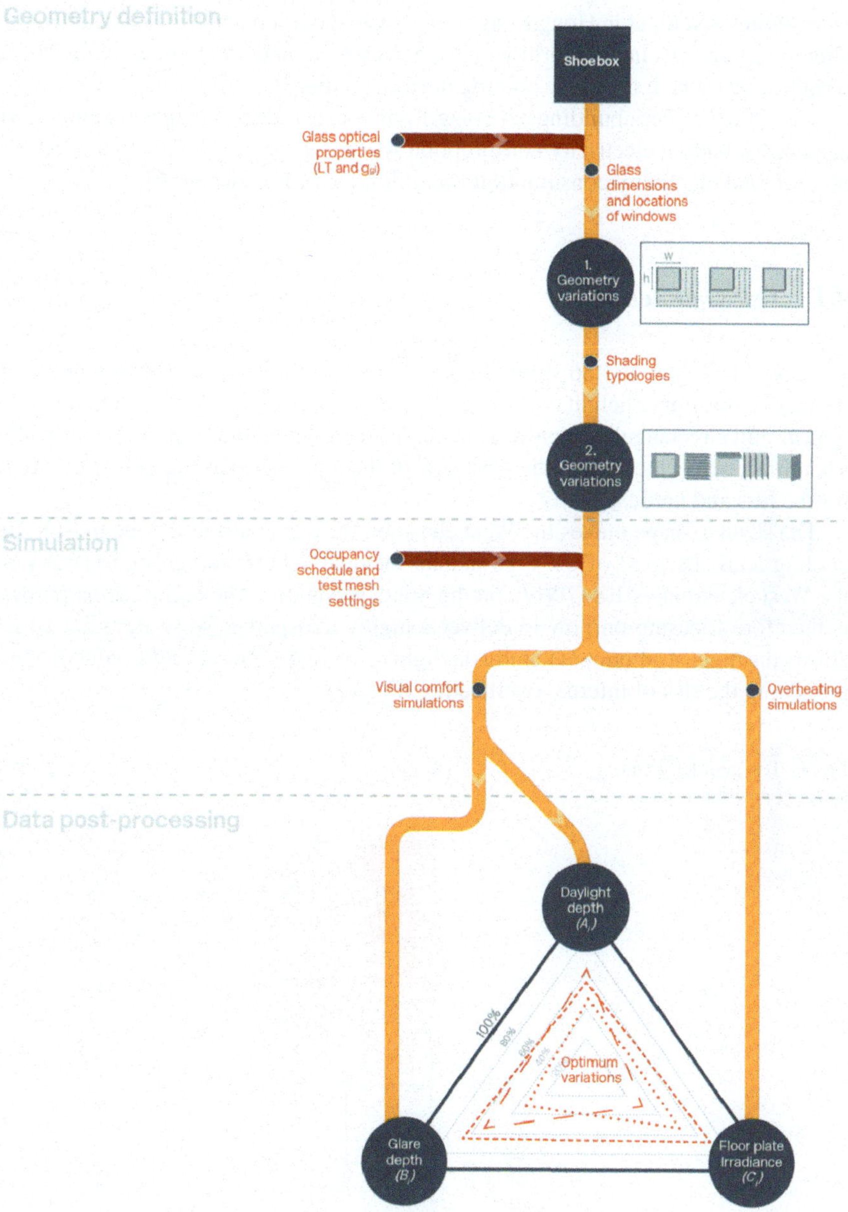

Fig. 5 Shoebox modelling workflow for visual comfort optimization and overheating mitigation

Building envelopes in Hong Kong need to withstand some of the most challenging climatic conditions in the world with high average temperature given by the "heat-island" effect which make the cooling needs, significant.

Therefore, modern buildings in Hong Kong must be able to reduce cooling needs and corresponding electricity consumption which typically accounts to around 55–60% of total electricity consumption for buildings in Honk Kong [6].

4.1 Test Geometry

The tower is 220 m and 36 storey-high and it is divided into three different blocks throughout the entire height of the building. Each block consists of 12 floors. It has a typical office rectangular open-plan layout (dimensions: 20m deep and 72m wide). The office area is located on the west side of the plan, whereas the east side is used for the lifts and vertical shafts.

The slender shape makes the West elevation the one contributing the most to the cooling loads because of the high annual incident solar radiation (4.5 GWh/y on the West elevation vs 1.8 GWh/y on the South elevation). The optimisation process is therefore focusing on how to deliver a highly transparent West elevation which allows the highest amount of natural daylight to get in the internal environment while mitigating the risk of internal overheating (Fig. 6).

Fig. 6 Test geometry and shoebox position

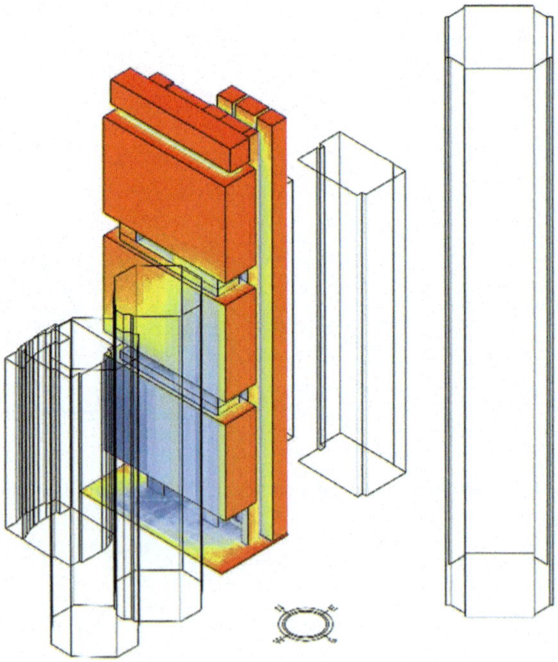

4.2 Input Variables

The workflow has been tested on shoebox-models in the middle of the highest block of the tower (with the highest incident solar radiation) with different facade options, including horizontal fins, vertical fins, and the articulation of the facade (rotation of the units around one or both middle axes).

The several facade iterations which had been collected and tested were created by setting the input variables of each shading strategy in its most extremes conditions and the increment step of each variable. As a design choice it has been assumed a fixed window-to-wall ratio of 65% to make the different shading strategies comparable (floor-to-ceiling height: 3.5m; window height: 2.5m; window-head-height: 3.2m; 10% reduction coefficient to take frame into account (Table 2).

$$WWr = 90\% * \left(\frac{2.5m}{3.5m}\right) \sim 65\%$$

The different geometries have then been tested with different glass optical properties, which differ in their selectivity index (LT = 30%, 45% and 60%; ggl = 0.3).

Because of each geometry being tested with the three different glass options, the total number of shoeboxes to be modelled and tested is equal to 37 (25 facade articulations, 6 vertical fins and 6 horizontal fins).

The models have been tested with 25cm grid size test mesh being positioned at a height of 85cm for the daylight simulations and 10cm for the irradiance-on-the-floor-plate simulations.

As additional input for the daylight simulations only, an occupancy schedule has been applied to the models to simulate people being inside the room and potentially experiencing visual discomfort; rooms have been assumed to be occupied 7 days a

Table 2 Tested variables for different shading options

	Horizontal fins		Vertical fins		Facade articulation	
Tested variables	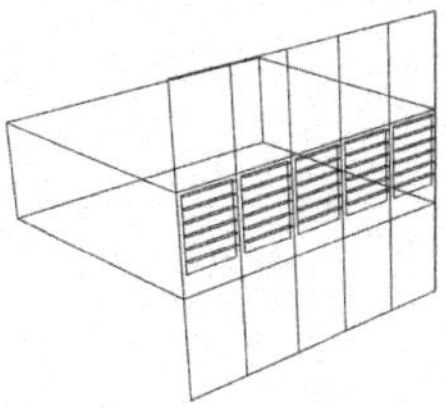		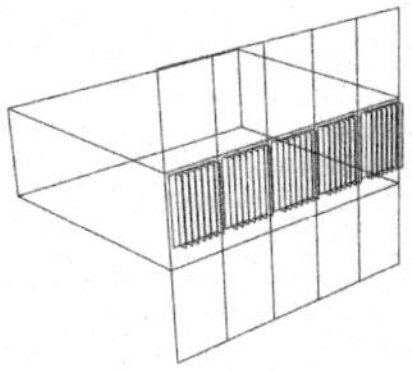		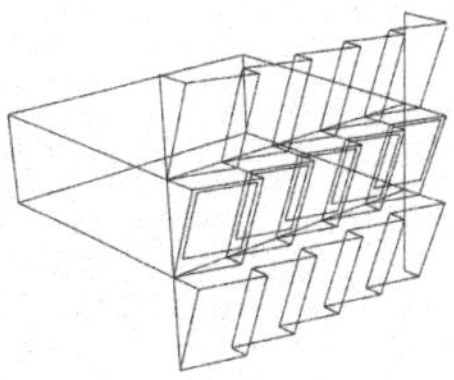	
	Fin depths	20 cm, 40 cm	Fin depths	20 cm, 40 cm	V rotation range	± 20° (5° intervals)
	N° of fins	3, 6, 9	N° of fins	3, 5, 7	H rotation range	± 20° (5° intervals)

week between 8am and 6pm with two hours lunch break from 12 to 2pm, as per the following chart.

4.3 Results

Table 3 shows an example of the three values (sDA depth, ASE depth and yearly average irradiance) collected for each one of the 37 shoebox-models.

All results have been normalised according to Sect. 3.3 to compare all the different facade options and to maximise the visual comfort conditions and reducing the risk of overheating.

Table 4 compares the three different passive strategies which have been investigated (external fins, horizontal fins, and facade articulation) with the baseline flat facade and shows the best one for each facade design as a combination of the best glass option with the optimal shading for visual comfort and overheating risk mitigation.

The baseline flat graph shows three curves corresponding to the three glass options, whose results are reported in the grey-filled cells and allow to compare the shaded facade options (blue, green, and purple-filled cells). The other iterations are divided into each shading strategy options and the thick lines show the optimal facade iteration for each of them (an additional image on the first row shows what the optimal configuration looks like).

The facade articulation shading option performs better with a darker glass (low LT) as it does not have any external additional shadings, whereas horizontal fins work better with a "dense layout" (9 fins, 30cm deep) with a more selective glass. Finally vertical fins require a highly selective glass with a low g-value to reduce the peak

Table 3 Tested variables for different shading options results from shoebox-models for multicriteria optimization

3 horizontal fins, 15 cm deep, LT = 45%, ggl = 0.3	
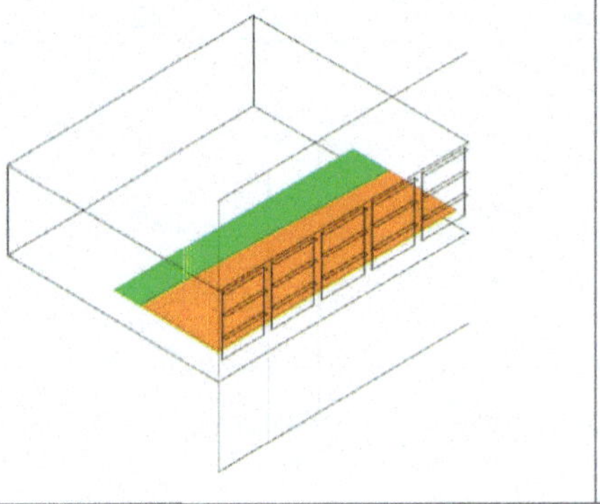	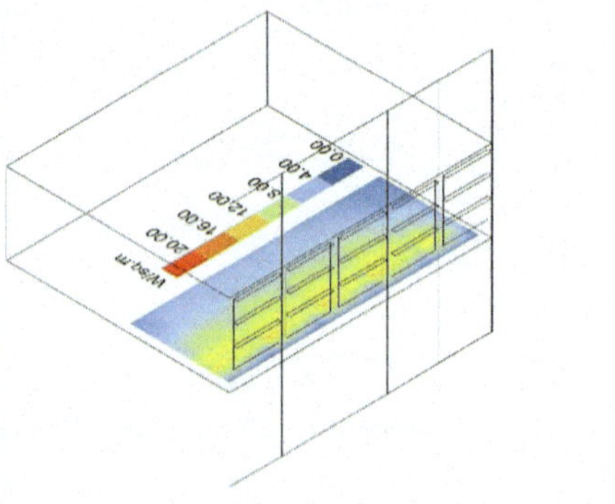
$sDAdepth = 4.65m$ $ASEdepth = 3.32m$	$G_{annualmean} = 6.2 W/m^2$

Table 4 Tested variables for different shading options results from shoebox-models for multicriteria optimization results for different passive designs

	Flat baseline	Facade articulation	Horizontal fins	Vertical fins
	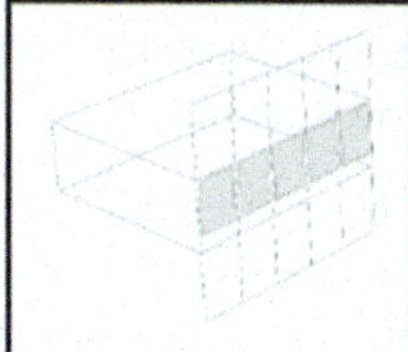	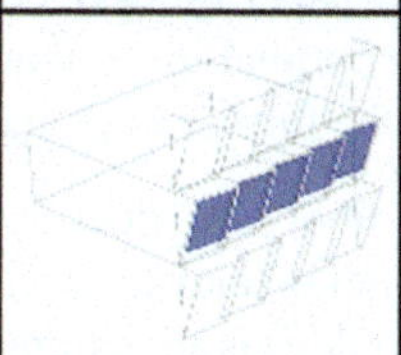 20° down | 10° North	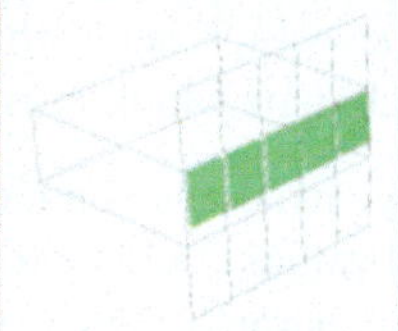 9 fins, 30 cm deep	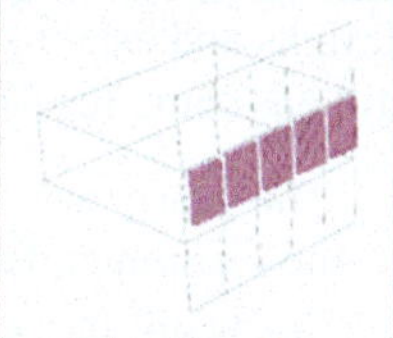 7 fins, 30 cm deep
Iterations	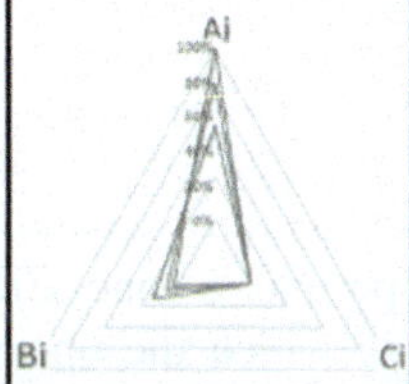	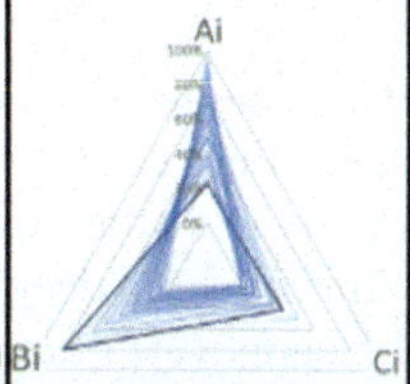	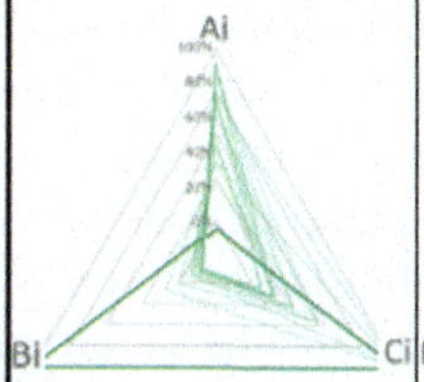	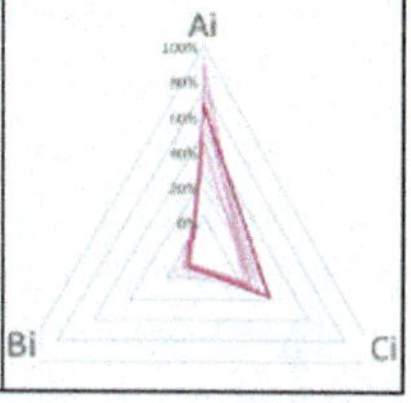
LT = 30% | g = 0.3	$sDA_{depth} = 3.68$m $ASE_{depth} = 2.75$m $G_{yearly\ av.} = 7.7$W/m^2	$sDA_{depth} = 2.19$m $ASE_{depth} = 1.16$m $G_{yearly\ av.} = 3.7$W/m^2	Non-optimal solutions	Non-optimal solutions
LT = 45% | g = 0.3	$sDA_{depth} = 4.86$m $ASE_{depth} = 3.52$m $G_{yearly\ av.} = 7.7$W/m^2	Non-optimal solutions	$sDA_{depth} = 2.75$m $ASE_{depth} = 0.92$m $G_{yearly\ av.} = 1.6$W/m^2	Non-optimal solutions
LT = 60% | g = 0.3	$sDA_{depth} = 5.81$m $ASE_{depth} = 4.16$m $G_{yearly\ av.} = 7.7$W/m^2	Non-optimal solutions	Non-optimal solutions	$sDA_{depth} = 4.17$m $ASE_{depth} = 3.09$m $G_{yearly\ av.} = 4.2$W/m^2

irradiance values (at peak hours the sun will be in front of the facade and vertical fins will not obstruct the sunrays).

Overall, the optimal facade iteration appears to be the one with horizontal fins (all the radar charts can be compared among themselves, and the thick green line is the one achieving the greatest triangle area value—as per "3.3 Data post-processing").

As additional requirement, to avoid an excessively dark room, a minimum sDA-depth has been set to two 2 m. To put that in context, in 2012 Christoph Reinhart has carried out some massing studies from which he managed to derive a set of rules of thumbs according to which a standard windowed facade can produce useful illumination to a depth of about 1.5 times the height of the window. This would lead to a maximum achievable daylight depth of approximately 4m.

5 Limitations of the Study and Future Outlooks

The workflow proposed in this research paper quickly compares different facade designs and establishes which one is the optimal in a comparative way. However, it still has some research gaps which must be further implemented; mainly:

1. The yearly average irradiance on the floor plate does not have thresholds set and no recommended values could be found in relevant documents. The risk of using this result in a comparative way is to design an extremely dark room or, on the other hand, a room which has overheating problems given that it exceeds the indoor comfort temperature anyway.

   This risk has been partially solved by running annual irradiance-on-the-floor plate simulations and checking how many times the floor-averaged value exceeds a certain threshold throughout the year. BCO's office specifications recommend not to exceed a threshold of 50-65 W/m^2 on 4.5m deep perimeter zone. CIBSE TM52 sets a limit for the number of hours (3%) during which the operative temperature can exceed the threshold comfort temperature by 1K or more during the occupied hours. In absence of further references, this limit is currently used and the scale 0h to 250h is defined under the assumption that a typical office building is occupied between 7am and 6pm, 5 days a week from BS ISO 17772-1:2017 Annex O (2868 hours a year); this implies that ideally the overheating threshold should not be exceeded for more than 86 hours in offices.
2. The workflow focuses on building envelopes in hot climates, where the cooling load can be drastically reduced by reducing the solar gains. Temperate and cold climates may benefit of solar gains during the heating season.

   By replacing the irradiance yearly mean value with the facade overall U-value it is possible to optimise the facade thermal performance in cold climates, where conduction losses are more severe and solar gains are more needed (hence reducing the floor plate irradiance would lead to an erroneous facade design).

   No internal blinds are currently included in the script. this kind of dynamic shadings are typically used to prevent glare issues and overheating discomfort at very extreme times of the year.

   Construction and building engineering current challenges are nowadays analysed at EOC with similar workflows able to process large datasets originated from hourly simulations throughout the year. Examples of these are raytracing analyses for caustic surfaces, glare in point simulations and shading design through façade energy models. These simulations capture the yearly performance of any building component which can be therefore designed in its most severe conditions to prevent harmful incident or discomfort events from happening.
3. The proposed workflow is based on the assumption that limiting solar gains is always the most efficient strategy to deliver a ultra-high performance facade. Although this may be true for most buildings in the tropical region, this is not always the only possible solution: as an example natural ventilation can partially mitigate an excess of heat in the internal rooms caused by a clearer glass. For this reason, Eckersley O'Callaghan is currently implementing a new version of

this workflow—LEAF (Low-Carbon, Efficient, Adaptive Facades)—which will be energy modelling-driven; LEAF aims to reduce even further the gap between MEP Engineering and Facade Engineering by calculating via energy models what the impact of a specific facade design would be in terms of energy needs and operational carbon.

6 Conclusions

Multi-criteria optimization workflows quickly compare different facade designs to establish the optimal one, able to provide optimal daylight comfort internal conditions and low risk of overheating due to excessive solar gains. Current protocols have well established approaches to assess daylight comfort in rooms, but no time efficient methodologies to assess overheating risk in internal environments.

Within this background, the novel workflow introduces a flexible and human-centred design approach for buildings at early-stage. Shoebox models allow the designer to include different glass properties (solar factor and light transmittance) and different shadings to be tested allowing the workflow to capture different glass-shadings build-ups (geometrical flexibility); this innovative approach gives the users the freedom to choose which occupancy schedule to use (building typology flexibility). Users can also include different material-specific optical properties and different simulation settings—Radiance parameters—to capture the effect of different finishes on the internal comfort conditions, such as highly reflective or translucent materials (material flexibility). The purpose of shoebox modelling is to be adaptable to different climatic files, which the user can set by assigning different EPW/WEA files in the inputs of the analysis (climatic flexibility).

While many papers had already captured how individually geometry, material, building typology affect the comfort of the building users, this workflow creates a link between all of them and gives the designers a human-centred tool to design ultra-high performance building envelopes for people's comfort and wellbeing. Future steps are to include additional parameters, such as the embodied carbon, U-values and eyesight lines from inside to outside to deliver even greener and comfortable solutions able to reduce the overall embodied carbon of ultra-high performance building envelopes.

References

1. Baker N, Steemers K (1999) Energy Systems (1st ed)
2. Building Research Establishment's Environmental Assessment Method (2018) Breeam UK New Construction: Non-Domestic buildings
3. Chartered Institution of Building Services Engineers (1987) Daylighting and window design
4. Chartered Institution of Building Services Engineers (2013) TM 52: The limits of thermal comfort: avoiding overheating in European buildings

5. Costanzo V, Evola G, Marletta L (2017) A review of daylighting strategies in schools: State of the art and expected future trends. In Buildings 7:2. MDPI AG. https://doi.org/10.3390/buildings7020041
6. Environment Bureau (2017) Hong Kong's climate action plan 2030+
7. Hamedani Z, Solgi E, Skates H, Hine T, Fernando R, Lyons J, Dupre K (2019) Visual discomfort and glare assessment in office environments: A review of light-induced physiological and perceptual responses. Build Environ. https://doi.org/10.1016/j.buildenv.2019.02.035
8. Hawkins G (2011) Rules of thumb guidelines for building services (5th Edition)
9. HM Government (2021) Approved Document O Overheating the building regulations 2010 Requirement O1: Overheating mitigation Regulations: 40B. www.gov.uk/guidance/building-regulations-
10. International WELL Building Institute (2016) The WELL building standard
11. Leadership in Energy and Environmental Design (2019) LEED v4 for building design and construction.
12. Méndez Echenagucia T, Capozzoli A, Cascone Y, Sassone M (2015) The early design stage of a building envelope: Multi-objective search through heating, cooling and lighting energy performance analysis. Appl Energy. https://doi.org/10.1016/j.apenergy.2015.04.090
13. Veitch JA (2002) Principles of healthy lighting: highlights of IE TC6–11's forthcoming report. www.lightingresearchoffice.org
14. VELUX Knowledge Centre for Daylight, Energy and Indoor Climate (DEIC) (2014) Daylight, energy and indoor climate basic book
15. Wargocki P, Wyon DP (2006) Effects of HVAC on student performance. www.ashrae.org
16. World Green Building Council (2019) Bringing embodied carbon upfront. www.worldgbc.org/embodied-carbon

Alessandro Baldini Associate Director at Eckersley O'Callaghan, since joining in 2014, Alessandro has been pivotal in the firm's growth—helping set up the Milan office which he now leads. With a strong multidisciplinary background, he has amassed extensive experience in facade engineering with a portfolio of clients and projects across the world. His approach to design is underpinned by a passion for sustainability and attention to technical detail, allowing him to operate successfully across different markets and sectors. Chartered as an Architect and Engineer in the UK (MCIBSE), he is also member of the Order of Engineers in Italy. Alessandro has published several research papers and presented at international facade and glass conferences. He currently leads the research and development stream for the practice's Facade Engineering group, promoting a strong focus on innovative construction technologies, circularity and low carbon facade design.

Claudia Brisolin a Project Engineer, joined EOC in December 2019 after completing her MSc in energy analysis and building system optimization. She specializes in designing complex facade systems for global projects, utilizing advanced software and extensive knowledge of materials and production methods. Claudia's expertise and capabilities make her a valuable asset for any project.

Emiliano Adam Cerutti a graduate facade engineer, joined EOC in 2021. His primary focus lies in implementing decarbonization strategies for EOC's projects throughout Europe and Asia. His key responsibilities involve the application of building physics and energy modelling principles to mitigate operational carbon emissions across various assets, all while ensuring optimal indoor visual and thermal comfort. Leveraging his expertise in Digital and Computational Design, he pioneers innovative methodologies for processing large-scale datasets using parametric tools to conduct early-stage design assessments.

Chapter 8
Testing Digital Workflows on Building Renovations: The Case Study of F10 New Law Building for the University of Sydney

Alberto Sangiorgio, Alex Bogdanova, Anastasia Globa, Catherine Jacob, Veronica Vela, and Mirco Bianchini

Abstract In recent years, digital tools and technologies like parametric design, computational modeling, and advanced manufacturing have disrupted the Architecture, Engineering, and Construction sector. Despite traditionally slow adoption, these tools have the potential to revolutionize building design and construction. While many case studies demonstrate the cost, energy, and carbon savings achieved through their application in new and large-scale projects, their potential for retrofitting existing buildings remains largely unexplored. This chapter presents a case study of a façade retrofit at the University of Sydney's 'F10—New Law Building' to showcase the benefits of digital technologies in analyzing, reconfiguring, and optimizing existing structures. The authors utilized various digital workflows to assess the façade's condition, evaluate design options, and involve stakeholders in identifying optimal solutions.

A. Sangiorgio (✉) · A. Bogdanova · M. Bianchini
Mott MacDonald, 383 Kent Street, Sydney, NSW 2000, Australia
e-mail: Alberto.sangiorgio@mottmac.com

A. Bogdanova
e-mail: Alex.bogdanova@mottmac.com

M. Bianchini
e-mail: Mirco.bianchini@mottmac.com

A. Globa
The University of Sydney, G04 Wilkinson Building, Camperdown, NSW 2006, Australia
e-mail: Anastasia.globa@sydney.edu.au

C. Jacob · V. Vela
The University of Sydney, G12 Services Building, Camperdown, NSW 2006, Australia
e-mail: Catherine.jacob@sydney.edu.au

V. Vela
e-mail: veronica.vela@sydney.edu.au

P. Ruttico (ed.), *Coding Architecture*, Digital Innovations in Architecture, Engineering and Construction, https://doi.org/10.1007/978-3-031-47913-7_8

Keywords Digital workflow · Façade retrofit · Shoebox analysis · Virtual reality · Computational modelling

1 Introduction—Digitalization of the AEC Industry

In recent years, several industrial sectors have undertaken a process of digitalization that involves the development and deployment of novel digital tools and digital workflows.

The Architecture, Engineering, and Construction (AEC) industry is recognized to be among the slowest sectors to adopt new changes, mainly due to its risk-averse nature. Nonetheless, it is witnessing a significant digital push across its various disciplines and areas towards the adoption of innovative workflows and tools [1].

Indeed, considering that the main drivers of innovation in the AEC industry are productivity, sustainability and creativity [2], digitalization is recognized to be the key enabler capable to crosscut these drivers and unlock significant benefits.

The use of digital workflows and tools is highly beneficial for the optimization of production processes that, ultimately, enable the reduction and management of resources towards more sustainable outcomes. Further, they enable design for manufacturing approaches that were, before, unthinkable, hence allowing designers and builders to imagine, obtain and manage forms and techniques that, in the recent past, would have been considered otherwise impossible.

In this context, the AEC sector is undertaking a slow yet steady transformation that diffuses across all different stages of the design and construction process. Thus, each and every player in the construction field is, on its own, dealing with digitalization.

A vast and growing pool of case studies demonstrates the economic and environmental benefits associated with the use of digital technologies for the design of new buildings. Despite these technologies being widely embraced by designers and builders for new building developments, currently their application for the retrofit of existing buildings is significantly underexplored, leaving the open question.

Can novel digital technologies be beneficial to the analysis, reconfiguration, and optimization of existing buildings?

Considering that 80% of the 2050 building stock is already in existence today [3], understanding how to optimally manage and renovate the existing buildings is key to achieving the net zero emissions target set for 2050, the associated high building performances and the related targets around the use and re-use of resources (ref. circular economy).

This chapter attempts to provide an answer, specifically looking at the renovation of the F10—New Law Building of the University of Sydney, located in Camperdown (Sydney, Australia). It outlines how existing digital technologies commonly used to address the design of new construction, can also be employed for retrofit interventions to, in this case, rapidly analyze the existing façade, evaluate multiple design options, and optimize the final design solution.

This project represents atest bench for several digital technologies. The workflow illustrated in this chapter was developed through a collaborative approach between Mott MacDonald and The University of Sydney (the University Infrastructure team and School of Architecture, Design and Planning), and it serves the purpose of identifying successful digital workflows that could be easily deployed on future and more complex projects.

2 Scope of Work

The case study presented in this chapter is the renovation of the F10—New Law Building at the Camperdown campus of the University of Sydney. Designed by FJMT Studio, this building was completed in 2009 and it has been a famous landmark for the campus and the city ever since. Thanks to its innovative spatial arrangement, aimed at maximizing engagement, collaboration and flexibility, and the use of a mixed mode ventilation with double-skin façade with integrated timber shading, this award-winning building has been featured in several publications as a state-of-the-art teaching and learning facility for the Sydney Law School [4, 5].

However, after the Grenfell Tower incident in 2017, the Lacrosse Building fire in Melbourne in 2014 and the more recent Neo 200 tower fire in Melbourne in 2019, the construction industry and policy makers made a fundamental reappraisal on the use of combustible materials in buildings and especially on facades, by tightening the interpretation of the existing building codes, and by introducing new and more stringent requirements concerning the fire safety of new buildings.

In New South Wales (NSW), Australia, the Environmental Planning & Assessment Regulation 2000 (EP&A Regulation) was amended to require the registration of certain existing buildings that have external combustible cladding (for Type A/Type B Buildings ref. Australian National Construction Code—NCC2022 C1.9 Non-combustible building elements).

The F10 building, designed in compliance with a previous version of the National Construction Code, employed combustible cladding panels and shading elements, including aluminum composite panels (ACP), engineered timber panels, and plywood solar shades (Fig. 1). With the advent of the new building codes (from NCC 2016—Amendment 1 onwards), it was identified by the University of Sydney as a priority building that needed to be rectified.

In 2022, following an initial assessment and subsequently issued Fire Safety Orders by the NSW Department of Planning, Industry and Environment (DPIE), the University of Sydney commenced the replacement of the combustible cladding on several buildings across their Camperdown/Darlington campus, including the F10 building.

The F10—New Law Building's Facade

The F10—New Law Building comprises five levels above ground and consists of a teaching and research facility for the Sydney Law School. Situated on a narrow site

Fig. 1 The F10—New Law Building, the University of Sydney campus (Sydney, NSW). The building, completed in 2009 was designed by FJMT Studio (today Fjc Studio)

on the campus, the building presents an elongated shape, spanning a north–south axis. Consequently, most of the surface of the facades, approximately 8000 m^2, has a predominant east and west orientation (Fig. 4), facing the iconic Quadrangle building on one side, and Victoria Park on the other.

To manage the high solar radiation loads hitting this surface whilst maintaining the visual connection to the outdoors, the façade is a fully glazed double skin with operable solar shades placed within the 1130 mm ventilated cavity. The shading system consists of vertical fins equally spaced along the whole façade. Visually, the fins present a consistent and uniform aesthetic, however, they are grouped into arrays. In correspondence of each bay (meaning the space between columns) the external shading system consists of a lower array of small fins 1110 mm tall, and an upper array of larger fins 2190 mm tall (Fig. 2). Each array is composed of five fins and is automated and separately operated by the building management system with manual override.

This double skin façade is interrupted only in correspondence with the glazed 'bridge' that connects the northern and southern areas, hosting circulation and informal collaboration spaces, and featuring a single-glazed skin (Fig. 3).

At the time of the renovation work, most of the existing vertical shades were reported as non-functioning, due to a failure of the mechanical components. As a result, most of the offices offered a sub-optimal indoor environmental quality, characterized by either insufficient daylight and view out, or excessive glare and overheating. The replacement of the combustible fins with a non-combustible alternative was highlighted as an opportunity to address these issues.

Operable automated shading systems are often preferred to fixed ones due to their ability to respond to multiple requirements and adapt to variable external conditions.

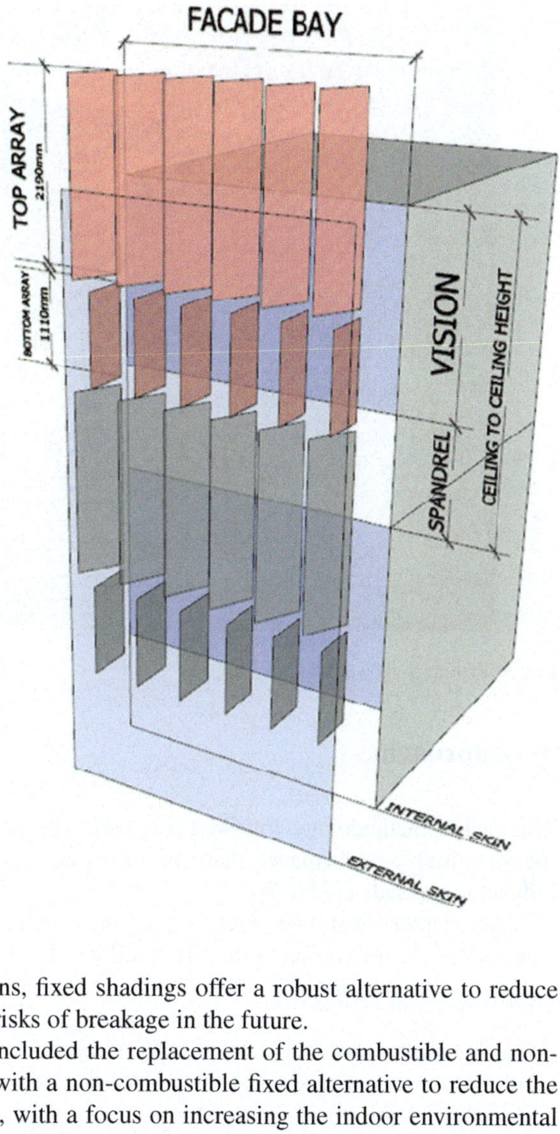

Fig. 2 Diagram of typical bay—double skin façade with vertical shading

However, in some applications, fixed shadings offer a robust alternative to reduce maintenance costs and limit risks of breakage in the future.

The final scope of work included the replacement of the combustible and non-functioning shading system with a non-combustible fixed alternative to reduce the inherent cost of maintenance, with a focus on increasing the indoor environmental quality experienced by users. In this instance, the design challenge was represented by the necessity of optimizing the shading orientation to balance competing parameters such as view, daylight, glare, and solar radiation (Fig. 4).

Mott MacDonald was appointed in early 2022 to assist The University of Sydney—University Infrastructure (USYD-UI) in responding to this design brief. The design team (USYD-UI and Mott MacDonald) partnered with the School of Architecture, Design and Planning at the University of Sydney to explore the integration of digital workflows with the decision-making process.

Fig. 3 Picture of the vertical fins

3 Approach

The design methodology followed a holistic approach aimed at optimizing both the quantitative performance (hard performance) and the qualitative aspects (soft indicators), as indicated in Fig. 5.

This chapter illustrates some of the steps undertaken to identify the optimal solution for the renovation of the F10 building façade, which included:

User perception analysis:

- Engagement with the existing building occupants to gather a greater understanding of the use of the space and establish design indicators that reflect the occupants' preferences/priorities.
- Identification of a set of design criteria, including the "hard" performance requirements defined by the relevant standard and the "soft" performance indicators indicated by the existing building occupants.

Desktop performance studies:

- Parametric modelling and shoebox analysis of the various design options.
- Multi-variable optimization to identify a final set of possible alternative design solutions.

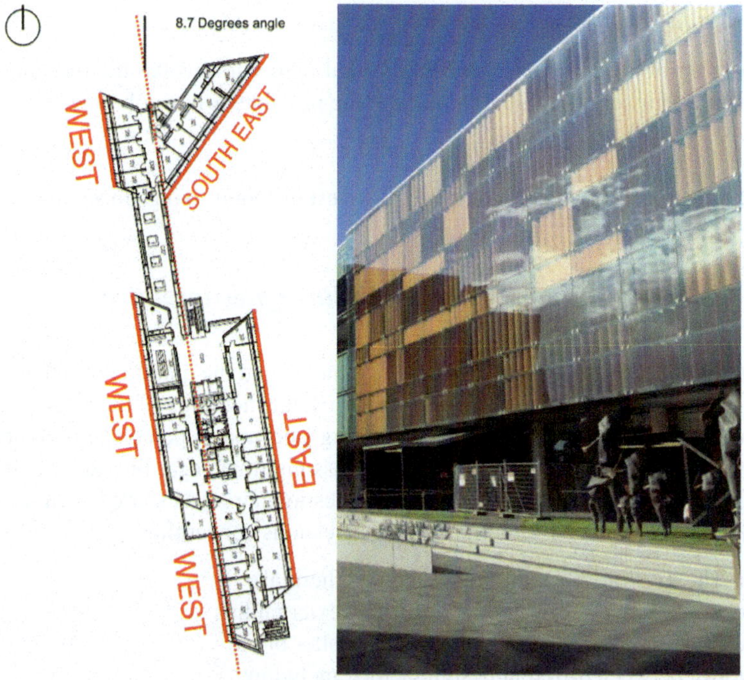

Fig. 4 Building orientation and location of double skin facades

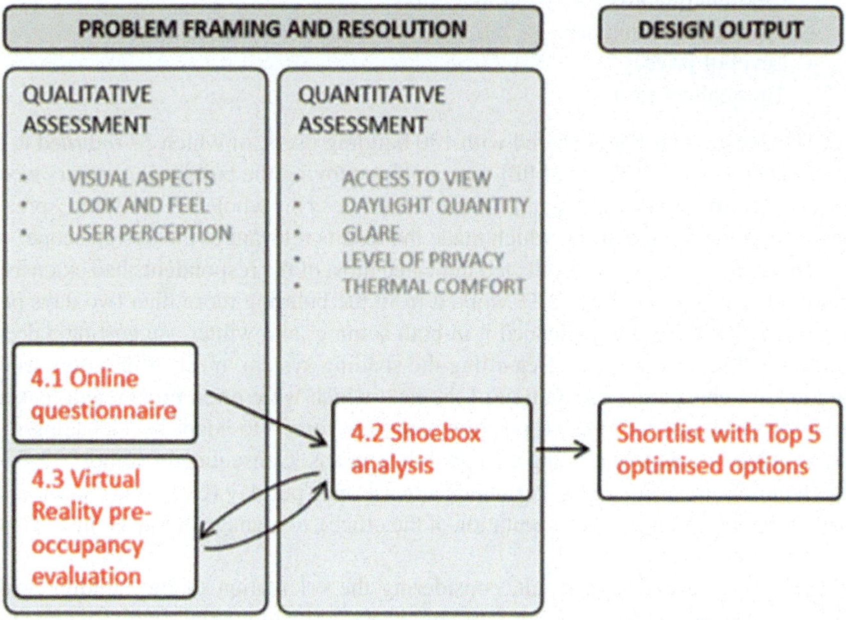

Fig. 5 Methodology diagram—4 steps digital workflow

Virtual reality applications:

- User-centric design approach using virtual reality to involve the relevant stakeholder and assess the visual impacts of the possible design solutions.

Early-stage prototyping through 3D printing:

- 3D printing of the façade components to test different fixings mechanisms.

3.1 User Perception Analysis—Online Questionnaire

The users' perception analysis aimed to (1) understand the qualities and flaws of the existing indoor spaces, (2) identify and establish the hierarchy of the relevant design criteria according to what the existing building's occupants perceive as important in their workplace, and (3) validate the design assumptions regarding the use of space and shading systems. To do so, an online questionnaire was shared with the target population, containing 14 multiple-choice questions concerning:

- Current level of user comfort (visual and thermal)
- Frequency/reason for use of the operable external fins
- Frequency/reason for use of the internal roller blind
- Priority between different design factors, including:
  - Access to view
  - Daylight quantity
  - Protection from glare
  - Level of privacy
  - Thermal comfort

The questionnaire was shared with 130 building users, of which 37 returned it—equivalent to 28% of the total full-time staff occupying the building. Answers were received from staff occupying different offices across the whole F10 building, spread in elevation and orientation, which made the results relevant to the design scope.

The analysis of the data collected indicated most of the respondents had extensive experience of the building, 97% worked from the building more than two days per week, and 100% had experienced it in both summer and winter, suggesting a deep understanding of the space. Regarding the shading system, most of the users were not able to change the orientation of the fins, which were often broken and locked in a too closed or too opened position and were forced to adapt by managing the solar radiation and overheating through other means. Consequently, almost 50% of the building users utilized roller blinds at least once per day (early morning or late afternoon depending on the orientation of the office), to manage thermal comfort and daylight intensity.

This was an expected result, considering the orientation of the building with prevalent elevations facing East/West, and it underlined the opportunity to increase the thermal comfort and quality of the entering daylight while replacing the existing combustible fins with a set of new fixed non-combustible fins.

The results were also used to identify the design priorities as perceived by the users. The main findings unveiled that:

- The level of privacy in the working space was not perceived as a high priority, whilst the average user valued access to view and daylight the most.
- The user satisfaction was low for 'access to view' and 'thermal comfort'.
- The east facing façade was associated with lower levels of thermal comfort satisfaction, whilst the west façade to lower levels of satisfaction in regard to view out, daylight and glare.

These findings suggested that the offices exposed to the West were more subjected to poor visual comfort conditions, while East-facing offices suffered from overheating, indicating that the design should focus on different priorities on the different orientations. Further, the results supported the ranking of the design factors based on the user's perception (Fig. 6), informing the multivariable optimization process undertaken at the back of the shoebox parametric modelling.

It is noted that the questionnaire addressed both the personal offices and the collaborative spaces, obtaining similar results.

3.2 Desktop Performance Studies—Shoebox Analysis

The scope of the shoebox analysis was to model a pool of design options through parametric modelling, evaluate their performance against the relevant parameters, and short-list the best-performing ones for the University to choose from to compose the final design of the facade.

The multivariable optimization was performed within the Rhino + Grasshopper suite, whilst the definition of the shoebox parametric modelling included:

- The identification of a set of variables, to establish all possible design options.
- The selection of project-specific parameters to assess the performance of each design option.
- The identification of control areas and reference offices.
- Scripting extracting the results using parametric methods.

Design parameters. The findings of the qualitative analysis informed the definition of the design parameters at play in the multivariable optimization, which was necessary to balance the competing design criteria. Specifically, the design aimed at:

1. Maximize outside visual connection (sky view factor)—Competing with points 2 and 4 below
2. Minimize solar heat gains to reduce overheating, especially on the East façade—Competing with points 1 and 3

Q9 – Using the scale below, please use the sliders to rate how much these aspects of your working space are important to you.

RANK	FIELD	MEAN	Std DEV.
1	DAYLIGHT QUANTITY	8.44	1.54
2	ACCESS TO VIEW	7.83	1.85
3	THERMAL COMFORT	7.39	2.41
4	PROTECTION FROM GLARE	7.25	2.63
5	LEVEL OF PRIVACY	5.85	2.85

Data broken down by elevation:

ELEVATION	DAYLIGHT	VIEW	THERMAL C.	GLARE	PRIVACY
EAST	8.75	8.87	7.56	7.56	6.42
SOUTH EAST	7.00	7.00	9.50	7.00	6.50
WEST	8.25	7.25	6.93	7.00	5.0

Fig. 6 Example of the information received from the questionnaire and subsequent data analysis

3. Maximize daylight—Competing with points 2 and 4
4. Minimize glare—Competing with points 1 and 3

These priorities were selected following a user-centric approach tailored to the building's existing population and informed by the questionnaire.

Based on these design goals, the design parameters were then further prioritized by favoring those that depended only on the design of the shading system over those that could be mitigated with other strategies. For example, the visual connection to the outside is a direct function of the external shading, which needs to be sufficiently open to allow the view outside, and, once the orientations of the fins are set, there are no other means to improve the view. On the contrary, the occurrence of glare within the internal space depends on the external shadings, but it can also be further mitigated by using internal roller blinds.

Establishing these priorities and agreeing with the ranking with the University was a fundamental step to resolving the complex design challenge and addressing the competing nature of the design parameters in the final assessment stage.

Geometric variables. The F10—New Law Building has been an iconic landmark for the University ever since its completion, and the design brief required the new fixed fins to maintain the same geometrical principles of the original fins regarding dimensions, spacing and proportion. This represented a significant design challenge, as it limited the optimization process, and the final optioneering exercise focused only on the rotation of the top and bottom fins.

The original rotation mechanism allowed the vertical shading elements to rotate from 0 (perpendicular to the facade) to −90° (parallel to the façade). This limitation on the rotation angle of the fins did not affect the efficiency of the shading system in its original conception as a mechanical movable system benefited from a limited motion range, however, including positive and negative orientations (−90 to +90°), demonstrated to be a much more efficient solution, especially on the West façade.

To reduce the number of computations, the final range of options excluded rotations above 60° and below −60°, as already known to be inefficient (too closed to allow a sufficient level of view and daylight), whilst considering steps of 10° rotations from the −60/+60 angle range (Table 1), which resulted in 13 possible orientations.

Shoebox analysis. Considering the scope of the shoebox analysis, a whole building simulation was deemed too resource-intensive and fine-grained, offering a level of detail unnecessary for the design objectives. Hence, the analysis focused on nine typical rooms, representative of the main offices and collaboration spaces in the building (Fig. 8) and the results of the analysis extended to the relevant areas on the elevation best represented by each room.

The selection of the rooms considered factors including:

- Orientation (W/E/SE);
- Overshadowing from both the F10 itself and nearby buildings;
- Size of the room, accounting for the differences between single offices and collaborative spaces.

It is noted that the analysis of all data related to the existing mechanical services was not available and the investigations related to the thermal comfort aspects were limited to the analysis of the solar radiation on the internal vertical surface by comparing the value of these parameters (Rad_m2 and Rad_summer, refer to Fig. 6) to a maximum solar radiation load established as a maximum baseline for each orientation.

The baseline was calculated by undertaking a survey of the existing façade.

All 9 rooms were modelled in the parametric computational design tool selected, and the proposed combinations of fin rotations were analyzed. For each option, all established performance indicators were computed and summarized for comparison.

Table 1 Variables for parametric design—shading orientation

Original design as envisioned by FJMT studio	Proposed geometrical possible solution for the façade resolution
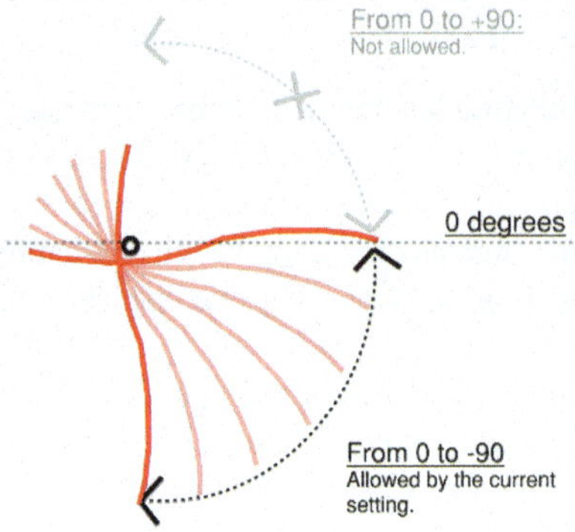	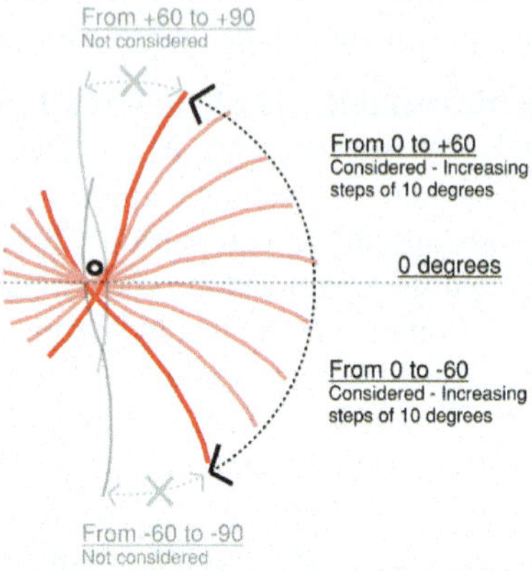
The existing design allowed the fins to rotate from 0 to −90° As a result of this limitation: – The shading capacity of the fins on the East and West elevations is different – In each elevation, all fins are oriented in the same direction (all negative angles). This results in a consistent look – On the West facade, the existing design does not provide an optimized orientation of the fins	The proposed design allowed the fins to rotate from −60 to +60° As a result of this increased range of options: – The fins orientation can be further optimized to achieve the best outcome – As a result of this scenario, the same elevation might host positive and negative orientations. The visual aspects will have to be further assessed to understand the impact on the aesthetic of the building

Considering 13 possible orientations (as per Table 1), 2 fins arrays (top and bottom, as per Fig. 2), and 9 rooms (Fig. 8), the total number of different resulting design combinations was 1521. For each combination, the 6 different design criteria described in Fig. 7 were quantified, resulting in a total of more than 9000 different results, as shown in Fig. 9.

All design solutions were then filtered by (1) excluding all those options that did not reach the established targets, (2) ranking the remaining options based on their performance on the criteria "view" and "solar radiation", and (3) shortlisting the five highest performing solutions that offered the best combination of orientations for the top and bottom fins (Fig. 10).

The shortlisted options were the top-performing design solutions that optimized the design criteria and priorities set based on the existing users of the building. Hence, it was important to assess the viability and acceptance of these options with the real users of the space. This step was performed in a virtual reality VR environment.

		PARAMETER	UNITS	TARGET	IMPORTANCE
VIEW	TOTAL EXTERNAL VISIBLE SURFACE	**Views_m²** External surface that is visible across the room	m^2	As high as possible	**Priority 1:** View was defined as high priority item by the building users.
		Views Percentage of external surface that is visible across the room	%	As high as possible	
THERMAL COMFORT	SOLAR RADIATION LOAD (WHOLE YEAR)	**Rad_m²** Cumulative radiation on the surface of the inner facade calculated for the whole year	kW/m^2	Equal or lower than the baseline	**Priority 1:** Thermal comfort is a competing parameter to view and the use of roller blind has a limited impact on it.
	SOLAR RADIATION LOAD (SUMMER ONLY)	**Rad_summer** Cumulative radiation on the surface of the inner facade calculated for the summer period.	kW/m^2	Equal or lower than the baseline	
DAYLIGHT QUALITY	DAYLIGHT ILLUMINANCE	**DI** % of area that comply with the requirements (min160lux /80% time - from Green Star Light Quality Credit).	% area	60-80% of analysed area	**Priority 2:** The illuminance of a room can be improved with artificial lighting
	ANNUAL SOLAR EXPOSURE (GLARE)	**Over** % of area that exceeds the requirements (max2000lux for 20% time - International best practices).	% area	< 20% of analysed area	**Priority 2:** Glare can be managed through the use of internal roller blind.
PRIVACY	NOT INCLUDED IN THE STUDY				**Priority 3:** Privacy was defined as a low priority item by the building users and can be managed though internal roller-blinds.

Fig. 7 Definition parameters for the shoebox analysis

3.3 Virtual Reality Application

User Perception and Design Evaluation Loop. The Virtual Reality (VR) application was introduced into this project to enable immersive exploration of the proposed façade system as a means to improve the design evaluation loop by meaningfully integrating both quantitative and qualitative evaluation metrics into the process. While several existing precedents from industry and research studies have previously used VR as a part of the pre-occupancy evaluation and the design/decision-making process, there is limited integration of performance simulation and qualitative design assessment in one holistic/combined VR solution [6–9].

Figure 11 summarizes the workflow that was developed to evaluate the user perception of the proposed design options in order to allow more informed collaborative decision-making within one combined package. After the qualitative analysis and the shoebox evaluation, the top-performing façade options were either accepted

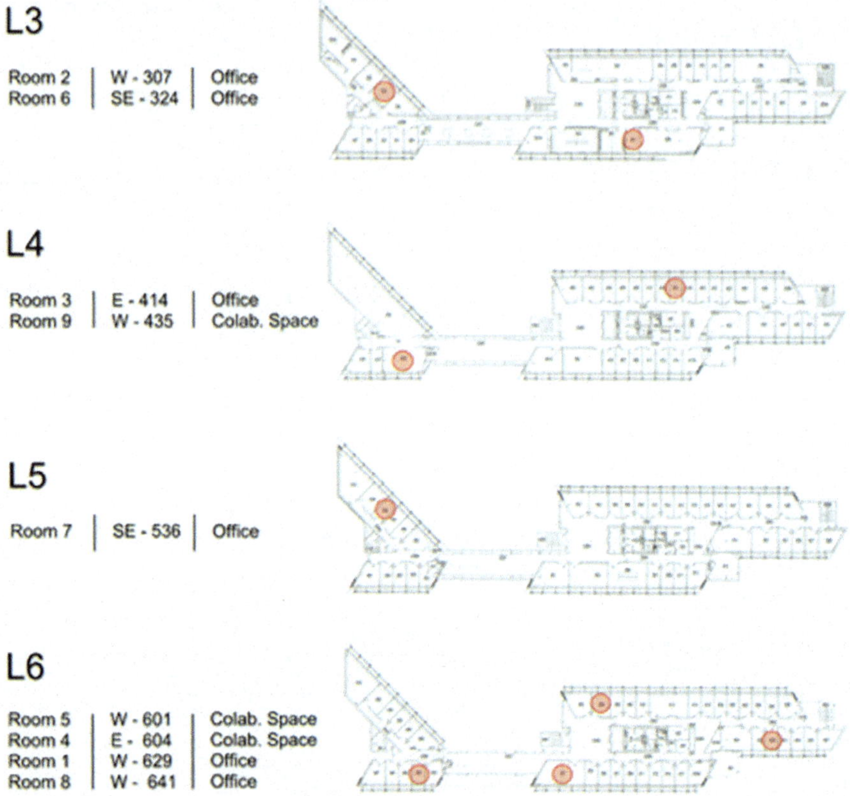

Fig. 8 Floorplans indicating the 9 control rooms representative of the whole facade

or rejected via the VR experience, not passing the qualitative assessment, triggering the next iteration of the design/evaluation loop.

The key metrics of the performance analysis (Fig. 7) were integrated into the VR User Interface (UI) and were previewed and dynamically changed when the user switched between the shortlisted design options.

The objective of the immersive VR environment was to allow the experience of both tangible and intangible aspects of office space where users could occupy the space through a range of dynamically changing external and internal conditions, engaging with visual, audio and kinesthetic human senses [10]. The functionality of the VR environment included the user's ability to:

- switch between shortlisted design options—assessing best-performing fin rotation configurations;
- rotate the fins at will to explore custom configurations outside of the shortlisted design range;
- change façade orientation to facing West, East, North, South or custom, which affected the views-out, the sun's position and sunlight direction;

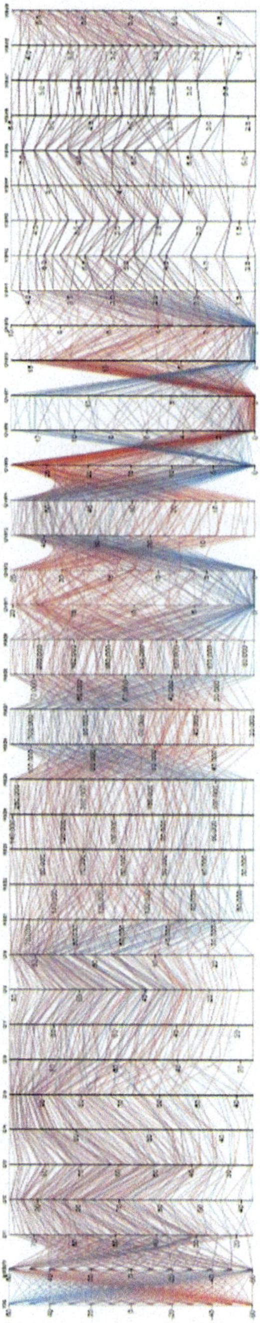

Fig. 9 Visual representation of the 9000 computations using the 'Design Explorer' tool

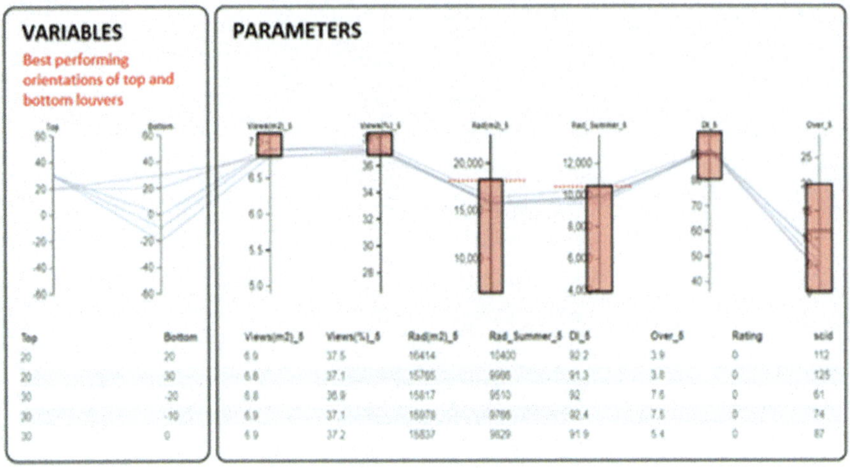

Fig. 10 Example of the output of the 'shoebox analysis' combined with the 'Design Explorer' tool

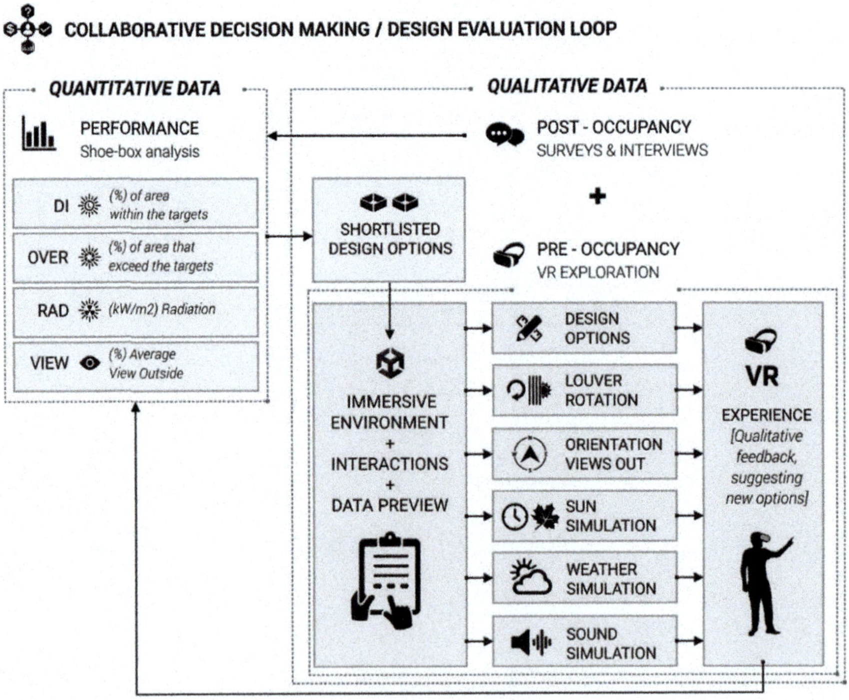

Fig. 11 Collaborative decision-making/design evaluation loop

- simulate sunlight and shadows, switching between the seasons (summer, autumn, winter and spring) and between the time of day (0–24 h);
- change weather between sunny, cloudy and rainy conditions.

To enhance the immersivity, the design team enabled the soundscape of the VR scene to dynamically change when the associated weather and time of day variables were activated. For example, the sound of the rain was activated when the user chose rainy weather. The daytime and night-time soundtracks were activated during and after working hours, respectively. Despite this feature having limited value for this specific project, it was tested in looking to future applications.

VR Development. Functionality, User Interface, and Interactions. Rhino3D (2022) was used to 3D model most of the interior spaces and façade system [11]. The VR scene development was done using a real-time development platform, 3D, 2D AR and VR engine—Unity3D (2022) [12]. The production pipeline consisted of two main parts: the development of the scene content (Fig. 12) and scripting behaviors and functionality (Fig. 13). Although the 3D model and textures of the existing office space could have been produced using photogrammetry or laser-scanning methods, we opted for using 3D modelling in Rhino to allow more control over the design of the interior space. The time and weather-specific soundtracks were recorded on-site and were further edited in the sound editing software.

The 3D geometry from Rhino was imported into Unity software as mesh objects (FBX format) optimized for the best performance and appearance in VR. All texturing

Fig. 12 Unity development

Fig. 13 Virtual interface

and visual effects, including the use of emissive materials for ceiling lamps and screen glow, was done directly in Unity. Small-scale interior scene objects, such as coffee cups, monitors, furniture, and books, were imported from Unity assets [13] and various online 3D repositories. The fin prefabs were imported separately from Rhino and were configured to have a design-specific pivot point to ensure accurate fin rotation as a part of the interactive façade system in VR (Fig. 12, left).

The Unity scene used mixed lighting settings that involved the back lighting of the overall interior space (static) and real-time light and shadows generated by the sun (dynamic). The sun simulation was achieved using SunLight [14]—a location-based time-of-day tool. To realistically portray various times of day, seasons and weather conditions, the range of the post-processing behaviours was applied to the VR camera (Fig. 12, right). The views outside of the office window played an important role in the qualitative assessment. It was important to create a realistic, believable representation of dynamically changing views without compromising the system's performance. A semi-transparent 360-view image map was used to avoid the need to 3D model the outside scenery of the office space. It was applied as a texture onto a sphere that was placed around the office 3D model (Fig. 12, bottom).

The design of the virtual interface and all functional UI tabs are illustrated in Fig. 13. The UI canvas was attached to the left-hand controller while the right-hand controller was used to 'press' the UI buttons via laser pointer and the index finger trigger. The night-time and day-time UI versions were developed to better suit different light conditions in the scene throughout different times of the day. The compact UI view could be enabled to reduce unnecessary visual distractions during VR exploration. The fins rotation, façade orientation and weather control tabs could be expanded as needed to avoid over-clutter of the main UI space.

VR Environment, Stakeholder Engagement and Future Work. As a result, the developed VR prototype could successfully simulate a wide range of design options, weather, and light conditions, as well as allow users to compare performance analysis data combining the quantitative and qualitative aspects of design evaluation (Fig. 13). The VR prototype was used in practice as a part of the real-world decision-making process of the fins façade development. Two collaborative workshops/decision-making sessions were conducted in November 2022 targeting two different stakeholder groups. Each session took about two hours—that included iterative VR experiences and follow-up discussions. The first workshop session involved the project managers and the architects. The second session involved senior members of the School of Law at the University of Sydney and the academic and administrative staff that currently occupy the building. All the above stakeholders had an integral role in the decision-making for the façade re-design project (Fig. 14).

The VR environment was output into two Meta Quest 2 headsets [15], with one of the headsets projecting its VR view into the large screen. This allowed all workshop participants to actively engage in the process and meaningfully contribute to the discussion (Fig. 15). The feedback from the involved stakeholders who used the VR prototype was overwhelmingly positive. For each of the East-facing rooms, three of the shortlisted design options were eliminated from the selection after the stakeholders experienced them in VR. Most of the participants felt that two of these options were blocking the views out of the window, and one design option was eliminated as visually not compelling. This further proves the value and contribution of immersive VR experiences for architectural and engineering decision-making and design processes as they can successfully communicate the intangible aspects of a place [16].

Future work for the VR prototype would involve targeted workshops that will engage with the larger cohorts of current occupants of the Law building and further work with the architectural design and construction teams. The objective of these workshops will be two-fold. Firstly, the feedback gathered from current and future occupants, architects, engineers and builders will be used to inform the final design proposal for the façade system. Secondly, the post-use survey will help gather more data on how the VR prototype can be improved in the future, particularly by looking into the extended functionality, interactions, and more holistic immersive experiences. For example, the next iteration of the VR prototype could involve the added ability to change interior finishes, fins design and materials, and furniture choices and incorporate the biophilic design features into the scene.

Fig. 14 VR experience with interactive user interface customised for the project

Fig. 15 Design evaluation workshops—The VR tool was used to engage with the client, allowing the university management to part of the decision-making process

3.4 Early-Stage Prototyping Through 3D Printing

Beyond the orientation of the fins, the design brief included the re-design of the shading device (with non-combustible materials) and its connection to the existing façade members from the existing motorized system to a fixed one. However, the latter required a higher level of flexibility, as it needed to accommodate a set of more than 2000 new fins, oriented to different angles (steps of 10° with 11 possible orientations). This requirement offered a great opportunity to employ a design for manufacturing and assembly (DfMA) and design for maintainability and reliability (DfM/DfR) principles applied to mass customization.

In principle, the system's design requirements were:

- To respect the initial design principles by matching the original aesthetic of the fixings.
- To be compatible with the existing fixing elements (Fig. 16).
- To reduce the installation complexity and allow each fin to be easily installed in the correct orientation, as per design intent.
- To enable each fin to be manually rotated in a 0° configuration for maintenance and cleaning purposes.
- To provide the possibility of manually rotating each fin to a different angle, responding to future needs and requirements by offering a higher degree of flexibility of the system.

The starting point of the design process was the study of the design principles and components of the existing system which was designed to allow a dynamic rotation of the fins, through a mechanism fully integrated within the façade's structural members. Despite the inherent 'beauty' of the detail, the disassembly of some portion of the façade showed the systemic malfunctioning and failure of the mechanism was caused by the breakage of the cogged wheel used to transfer the movement from the motor to the fin's pinion (Fig. 16).

To align with the DfMA and mass customization aspects, the fins were designed as one single aluminum extrusion repeated across the whole façade, and the variability of the system was embedded in the design of the connections. To achieve the design principles established above, the complexity of the connections was resolved by:

- Allowing variable rotations by using a single variable component, easy to fabricate in different shapes (one shape corresponding to each possible orientation) and install depending on the required orientation. (option 2 in Fig. 17, where different shapes of the lock-in plate, in green, allow to vary the rotation of the fins)
- Allowing variable rotations by using the same components, designed to adapt to different layouts (options 1 and 3 in Fig. 17, where the repeated equally-shaped components can be arranged on-site in different ways to vary the rotation of the fins).

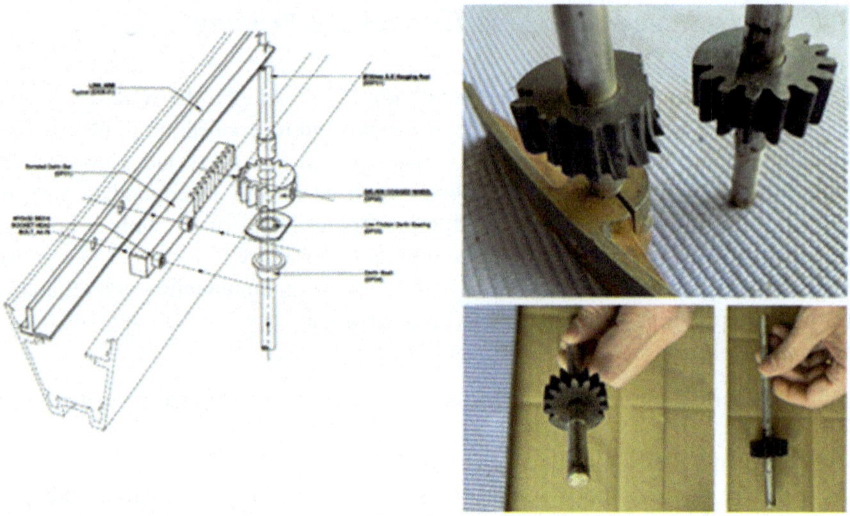

Fig. 16 Existing dynamic shading—failed teeth in the existing cogged wheel

This process produced several possible solutions, that are currently being explored through 3D printed prototyping, allowing engineers, fabricators and clients to experience and test first-hand the pros and cons of each system. The prototyping phase enables a thorough and integrated analysis of the different configurations beyond the aesthetic value, including different aspects of fabrications, installation, flexibility, and maintenance.

4 Conclusions and Future Outlooks

This chapter illustrated the digital tools and workflows for the renovation of the F10—New Law Building of the University of Sydney, used as a case study to demonstrate the applicability of this approach to retrofit projects. These digital technologies are more commonly used for the design and construction of new buildings, yet case studies on their deployment for renovation works are still scarce. Considering (1) the 2050 sustainability goals for the built environment, (2) the key role that digitalization plays in enabling its decarbonization, and (3) the fact that most of the 2050 building stock has already been built, retrofit and renovation of existing buildings is a compelling issue that can no longer be ignored. This chapter demonstrates that it is feasible to employ an integrated digital workflow in renovation works and it also unveiled the significant benefits that can be achieved by doing so.

Other than the obvious contribution to the project of shoebox analysis that enabled a data-driven decision-making process, the digital approach undertaken enabled the design team to engage with relevant stakeholders, both upper management and

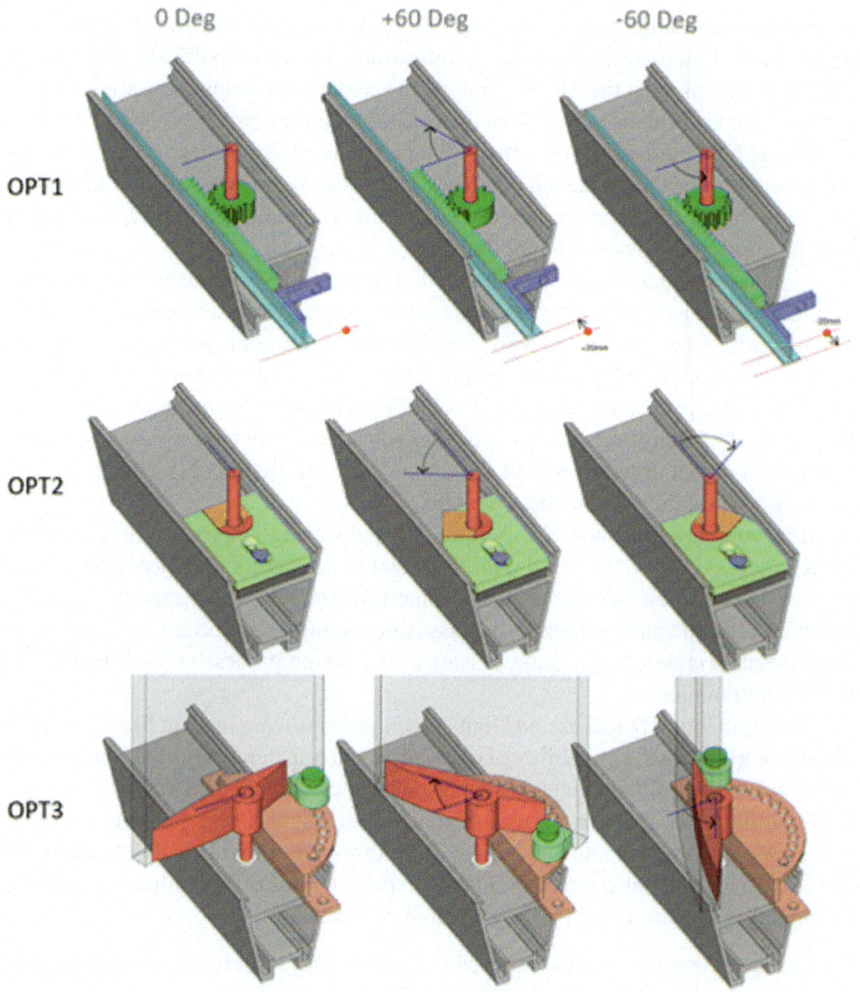

Fig. 17 Examples of the proposed design solutions subject to 3D printing and testing

everyday building occupants, and identify those design criteria that were otherwise impossible to set. Both the qualitative analysis and the VR studies have brought significant non-quantifiable benefits that determined the success of the design stage. The VR studies, for example, enabled testing design solutions and simplifying communications between designers and stakeholders, but, as a side benefit, unlocked the reduction of the time, cost and resources spent to generate visual and performance mock-ups that, at scale, support the transition towards a leaner and less carbon-intensive AEC sector.

Digital technologies have enabled, in this case, an integrated user-centric performance-based design approach, informing and cross-pollinating the multi-variable analysis and the shortlisting of the successful solutions based on their quantitative performance, as well as their qualitative aspects.

Despite being a small project, the F10—New Law Building case study offered an opportunity to develop a design workflow that is easily transferable and scalable to other new-built or renovation projects, and further implemented to meet the project-specific conditions, risks and opportunities.

The pandemic accelerated exponentially the creation of innovative technologies for the AEC industry, throwing the sector into the digital mindset arena and requiring a change in how we frame and resolve problems. In this context, a growing number of projects push the adoption of cutting-edge digital tools with the expectation of delivering cheaper, better, and faster solutions. Adopting or developing tools can help in the short-term, providing immediate convenience and improvements, yet it is essential to reflect upon the role of these tools and how the process should radically be transformed to maximize these benefits.

At Honda, they talk about the 'zero-based engineering' approach. They break the entire process down to the details and then rebuild it, looking for where things could go awry at each step, starting with raw materials through the finished product. The ability to examine and optimize a process is becoming as important as the ability to conceive an original design. Re-assessing every design process will, ultimately, aid to scale automation.

The Engineering Sciences and Digital Technologies team in Mott MacDonald are already working on the scalability of innovation and transferability of lessons learned across projects and disciplines, bringing more attention to standardized processes, and building tools that can be used repeatedly across different projects and practices. By codifying our knowledge into algorithms, Mott MacDonald can share not just information but also retain IP and the expertise built across projects. Capturing functionality in centralized platforms can improve:

1. Reusability and unification to develop standardization, and consequently increase quality control and reduce human errors.
2. Functionalities and workflows by continuing implementations in a more agile approach that consists of faster iteration in short test-and-refine loops.
3. Retention of knowledge and capabilities to maintain competitive advantages.

In an environment where the design team must work fast and promptly move from project to project, little time can be dedicated to the critical analysis of the work that has been developed and to the knowledge-capturing knowledge-sharing phases. To address and manage these fundamental aspects and improve our efficiency, we use dedicated strategies such as:

- Introducing specialized teams to 'productize' applications.
- Easy-to-use low code/node code tools. Reusable code libraries that are easily accessible.
- User-friendly catalogue system for asset 3d reusable.
- Platform as a service.

We believe that creating digital infrastructure to transfer ready-to-use and easily accessible workflows, nodes, and digital assets, is the key to unlocking a truly integrated and knowledge-based design approach. It is in this environment that projects like F10—New Law Building for the university will be able to inform and influence the next iteration of future projects across the campus (for The University of Sydney) and across the business (for Mott MacDonald).

Acknowledgements Writing this book chapter could not have been possible without the contribution to the project provided by:

Contributions to the Project: Ben Daykin (Façade Lead, Mott MacDonald—Project Principal), Peter Mellow (Project Director, The University of Sydney—Project Director), Risto Kotevski (Project Manager, The University of Sydney), Shivani Amin and Henry Wong (Senior Façade Engineer, Mott MacDonald—Design team).

Contribution to VR Experiment: Muhammed Yildirim and Kun Lyu (PhD Candidate, The University of Sydney).

Chapter Academic Review: Dr. Arianna Brambilla (A/Prof., The University of Sydney).

Chapter Final Review: Dr. Anne Hellstedt (Technical Excellence Leader—Australia and APNA, Mott MacDonald).

References

1. Gasparri E, Brambilla A, Lobaccaro G, Goia F, Andaloro A, Sangiorgio A (2022) Façade innovation: between 'product' and 'process'. In: Rethinking building skins. Woodhead Publishing, pp 1–13
2. Sangiorgio A (2022) Facade innovation: an industry perspective. In: Rethinking building skins. Woodhead Publishing, pp 15–58
3. Call for action: Seizing the decarbonization opportunity in construction. https://www.mckinsey.com/industries/engineering-construction-and-building-materials/our-insights/call-for-action-seizing-the-decarbonization-opportunity-in-construction. Accessed 25 Jan 2023
4. Morehen Thorp F-J (firm), Francis-Jones R, Morehen J, Thorp R, Morehen Thorp F-J (Firm) (2009) In the realm of learning: the University of Sydney's new Law School: incorporating the Faculty of Law, Law Library and General Teaching Complex, Images Publishing
5. FJMT Awards page. https://fjmtstudio.com/studio/awards/. Accessed 25 Jan 2023
6. Schiavi B, Havard V, Beddiar K, Baudry D (2022) BIM data flow architecture with AR/VR technologies: use cases in architecture, engineering and construction. Autom Constr 134:104054
7. Khalili A (2021) An XML-based approach for geo-semantic data exchange from BIM to VR applications. Autom Constr 121:103425
8. Tseng KC (2022) A feasibility study of using virtual reality as a pre-occupancy evaluation tool for the elderly. Autom Constr 134:104037
9. Globa A, Wang R, Tokede O, Khoo CK (2022) Pre-Occupancy evaluation of buildings in VR: development of the prototype and user studies. Archit Sci Rev 65(5):407–419
10. Moloney J, Globa A, Wang R, Khoo C (2020) Principles for the application of mixed reality as pre-occupancy evaluation tools (P-OET) at the early design stages. Archit Sci Rev 63(5):441–450
11. Rhino Rhinoceros 3D. https://www.rhino3d.com/. Accessed 5 Dec 2022
12. Unity Real-time development platform, 3D, 2D AR and VR engine. https://unity.com/. Accessed 25 Dec 2022

13. Unity Asset Store (2022) The best assets for game making. https://assetstore.unity.com/. Accessed 5 Dec 2022
14. SunLight (2022) Location based Time of Day. https://assetstore.unity.com/packages/tools/particles-effects/sunlight-location-based-time-of-day-66399. Accessed 5 Dec 2022. Meta Quest 2 (2022) Immersive all-in-one VR headset. https://www.meta.com/au/quest/products/quest-2/. Accessed 5 Dec 2022
15. Globa A, Beza BB, Wang R (2022) Towards multi-sensory design: placemaking through immersive environments-evaluation of the approach. Exp Syst Appl 117614

Alberto Sangiorgio is Technical Director in Façade Engineering and Eng. Sciences Practice lead for the APNA region (Asia, Pacific, New Zealand, and Australia) at Mott MacDonald. With a more than a decade-long background in architecture, engineering and construction, he has contributed to renowned projects worldwide, focusing on optimizing high-performance envelopes.

Alex Bogdanova is a Principal Sustainability Analyst at Mott MacDonald's Sydney office, focusing on computational design tools for high-performance and sustainable solutions. With a background in architecture and design, she offers a unique perspective on integrated design solutions.

Anastasia Globa is a Senior lecturer in Computational Design and Advanced Manufacturing at The University of Sydney. Her expertise lies in computational design and interactive immersive systems. Anastasia actively collaborates with research labs and communities, fostering academic and industry partnerships.

Catherine Jacob is the Façade Technical Director at The University of Sydney, ensuring the integrity and quality of facades for new build and re-clad projects. With a successful start-up and general management background, she has led prestigious projects across Europe and the Middle East.

Veronica Vela a Project Manager at The University of Sydney, adeptly manages significant projects from conception to completion. Her architecture background and structured approach facilitate seamless delivery of project requirements.

Mirco Bianchini is a Principal Digital Advisor at Mott MacDonald with 15+ years of AEC experience. His expertise lies in optimizing design and fabrication processes, and he is currently the product owner in the development and upscaling of 'Nodebox', a Mott MacDonald multi-platform library of computational tools.

Chapter 9
Computational artwork to transform places

Chris Fox and Tommaso Pagani

Abstract Studio Chris Fox has worked on various projects, from small scale temporary installations to large scale urban landmark artworks. This chapter presents three of the most recent projects that have been designed and delivered using computational design processes and digital workflow. The computational design approach of the Studio Chris Fox team allows rapid testing to push the aesthetic boundaries of every artwork. At the same time, this workflow and skillset enable the team to adapt and optimize different materials for fabrication integrating with the required engineering specification. The projects have been fully designed and documented within a parametric framework developed by the Studio Chris Fox team. The developed framework allowed for exchange of data and information with engineers, consultants and contractors. The chapter also presents insights of the coordination challenges between different stakeholders for the delivery of complex artworks.

Keywords Parametric framework · Complex artworks · Placemaking · Transforming places

1 Introduction

Studio Chris Fox (SCF) was founded by artist Chris Fox with the aim of transforming the built environment, bridging the disciplines of art, architecture and engineering. Chris is a senior lecturer at the University of Sydney, and has over 25 years' experience navigating the complex constraints of the public domain to design and deliver landmark interventions.

The Studio has a rich skill set that includes strategic art-thinking and computational processes that impacts urban design, architecture, landscape and place making.

C. Fox · T. Pagani (✉)
Studio Chris Fox, Gadigal Country, Sydney, NSW, Australia
e-mail: tommaso@chrisfox.com.au

C. Fox
e-mail: chris@chrisfox.com.au

P. Ruttico (ed.), *Coding Architecture*, Digital Innovations in Architecture, Engineering and Construction, https://doi.org/10.1007/978-3-031-47913-7_9

The studio transforms places with sculptural form that push the boundaries of what seems physically possible, whilst telling meaningful stories of place and people through time.

2 Projects

2.1 Rozelle Interchange

This landmark urban artwork interweaves between the Rozelle Interchange ventilation towers (Fig. 1). The form is derived from mappings of previous, current and future ecologies. These mappings generate patterns of sculptural looping and interweaving forms that provide a framework for nature to take over.

Situated on the land of both Gadigal and Wangal clans of the Dharug language group, three landscaped monoliths take ground above a new parkland on the edge of Rozelle Bay. The area that was once bursting with vibrant ecosystems has transformed over millennia; previously mudflats and mangrove forests, more recently characterized by industrialized maritime and rail use. This integrated landscaped artwork looks to this history and its continued stories.

Modulated zinc paneling and twisting steel structure (Fig. 2) references the turbulent air flow of the facility and projected spatial movements by inhabitants of the vast subterranean road network below. Enveloping the towers, this living system extends as bridge forms (Fig. 3) above pedestrians and cyclists, whilst biophilic green-wall

Fig. 1 Rozelle interchange, Sydney, Australia, 2018–2023, rendering

modules further integrate the structures into the parklands and transform this vast infrastructure into a habitat for urban biodiversity.

Fig. 2 Rozelle interchange zinc and green wall facade and bridging element

Fig. 3 Rozelle interchange bridging form from sampling platform

One of the first modelling processes of Rozelle Interchange was fluid dynamics simulation (CFD). The CFD model was used to inform the artwork form and the green wall panels modulation. To create a more organic artwork the CFD logic was replaced with slime mould and swarm agent simulation using Physarealm [1] and Quelea [2] plugins. The results were visually interesting but too complicated from the construction point of view. These workflows also did not allow for enough control over the zinc and green wall modulation.

A more manually controlled approach was then used to influence the design of the Rozelle Interchange artwork. This approach worked in parallel with the optimization procedures for structural performance, constructability and aesthetic outcome. A combination of Kangaroo [3], Galapagos [4] and Octopus [5] have been used to optimize the two bridges and the steel vines wrapping around the three stacks. Bollinger+Grohmann helped with the steel vines simplification. The collaboration and model exchange with them was supported by very detailed attributes assigned to every geometry in the Rhino® model. Geometry's attributes have been a key element throughout the entire project. This information helped translating the Rhino® geometry to Revit® for the documentation stage. Initially, GeometryGym [6] was used as a tool for interoperability between the two software. After a while it was replaced by Rhino.Inside®. Revit [7] working on Rhino® 7 WIP. This transition improved a lot the workflow in terms of time optimization due to live export of the tool. Moreover, including the Rhino® attributes in Revit® was more reliable and easier with Rhino.Inside®.

Steel vines and bridges are the key design elements of Rozelle Interchange. The two bridges are composed by straight and bent CHS (Circular Hollow Section). Initially, the steel vines connected to the bridges were welded but due to fabrication tolerance and accessibility, the elements have been detached. The steel vine connection to the bridge structure is now bolted. This upgrade also helped in reducing the transport dimensions of the bridging form. The maximum road allowed dimensions were provided by the steel fabricators and included in the modelling workflow. Several scripts helped dividing and checking that the bridge parts fit within the given numbers and oriented them on a semi-trailer to visually check any potential issue.

A similar approach was developed for the steel vines wrapping around the three stacks. Starting from the transport dimensions constraints, the elements have been divided in smaller parts with the biggest possible extension to fit in the bounding box. Most of the stitching elements were separated from the steel vines because they could not fit within the bounding box. Fig. 4 shows on the lowest part of the stack three vine assemblies with welded stitching elements that fit in the transport dimensions. The steel vine division was made by manually placing the split points.

Different test trying to automate the process as much as possible were made, but the amount of engineering and fabrication constraints made the process quite complicated and very time consuming given the design stage timeframe. The steel vines division script assigned (Fig. 5) a unique tag to the assemblies that was used for the shop drawing Tekla® model coordination and the fabrication stage. Moreover, live and dynamic check in the script have been put in place to ensure the assemblies did not exceed any design and engineering parameter.

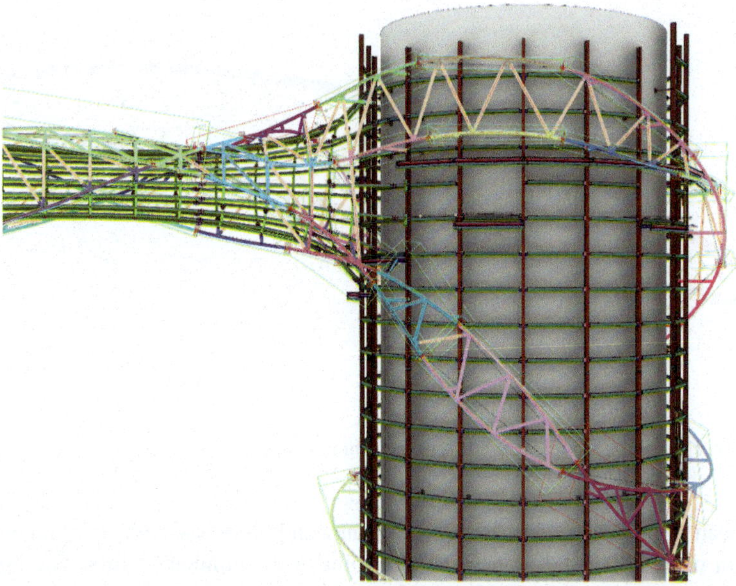

Fig. 4 Steel vine division for transport and fabrication, Rhino® model

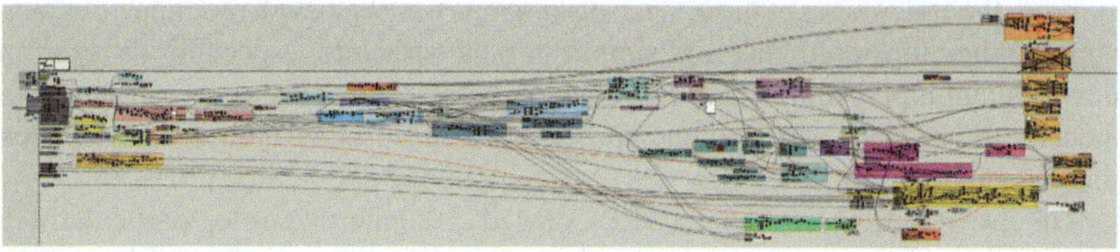

Fig. 5 Steel vine splitting script

Studio Chris Fox managed the shop drawing stage in collaboration with a shop drawing specialist consultant. All the required information (element tag, position on the structure, etc.) has been included in the Rhino® model attributes to inform the consultants how to proceed for the documentation. Close coordination with the steel fabricators also helped streamlining the required parameters and details for fabrication. Considering the complexity of the steel vines, additional drawings have been requested by the fabricators to check that every element was positioned correctly. The SCF team, starting from the steel vine division workflow and attributes, generated new scripts and models to show the vine assemblies on the fabrication jigs. For every assembly was created a reference drawing (Fig. 6), a fabrication file and a spreadsheet with coordinates to check the parts position in the workshop.

The whole structure was prefabricated, preassembled and painted off-site. On-site, the ready-made parts were only installed in their position following the general assembly drawings from the shop drawing specialist. Zinc and green wall modules

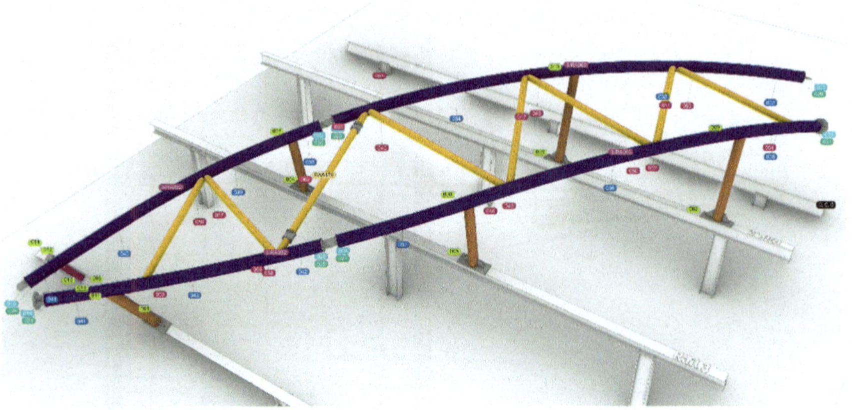

Fig. 6 Steel vine assemblies on fabrication jig, Rhino® model

have been preinstalled on columns and beams and all the assembled elements were lifted in place as a single big piece to minimize the number of lifts. The bridging forms have been preassembled and lifted on temporary supports before connecting them on the three stacks. The steel vines and their relative stitching elements were installed piece by piece due to the complexity of the geometry.

From the early stages of the project, discussions were held with the client to explore opportunities for optimizing and simplifying the structure, especially in regard to the installation procedure. This coordination helped identify areas where improvements could be made, streamlining the construction workflow and ensuring efficient implementation.

The Design Review Panel (DRP) played a supportive role, particularly in embracing the concept of nature taking over the infrastructure. The DRP provided valuable insights and inputs in this direction, reinforcing the project's vision and enhancing the integration of natural elements into the built environment.

Rozelle Interchange is a part of the WestConnex infrastructure project. The project team is composed by lots of different parties working on multiple areas and parts of the project. Constant coordination with the project design team was another key aspect of this project. The collaboration made sure that all the necessary information was included in the documentation and that the Revit® model was seamlessly integrated with the overall project model. Ideas and information exchange helped the design team in avoiding potential conflicts and delays between different design packages and guaranteed that the project could progress smoothly. Furthermore, the SCF project team engaged in model exchange with Bollinger+Grohmann engineers to optimize the structure. By sharing models and collaborating closely with the engineering team, design improvements and structural enhancements were identified and incorporated into the project. This iterative process helped create a more efficient and robust structure.

During the design development stage, discussions were also held with green wall suppliers. These conversations covered various aspects of the green wall system, such as plant selection, irrigation, and maintenance. By involving the suppliers early on, the SCF team was able to consider all the relevant factors and incorporate them into the design, ensuring a well-integrated and sustainable green wall system.

At the beginning of the shop drawing stage, coordination with the steel fabricator was crucial to include all the relevant information on the construction documentation. By providing comprehensive details to the fabricator, potential conflicts or issues during the fabrication and construction phase were minimized, enhancing the overall project efficiency. The detailed Rhino® model also helped in managing queries from the steel fabricators when needed. The streamlined modelling between the SCF and B+G team ensured additional information exchange for the steel vine fabrication. These exchanges allowed for continuous refinement and improvement of the fabrication process, ensuring that the steel vine was manufactured to the desired specifications and standards.

2.2 *Interchange Pavilion*

Interchange Pavilion (Fig. 7) is a meeting place where tracks converge, a place of interchange where paths cross. Drawing inspiration from the precinct's rail history, SCF worked with the salient geometries of the railroad switch; the point at which a train can change its course, moving from one trajectory to another. The work is a tribute to the diverse stories and journeys of people connected with the Eveleigh Rail Yards.

Interchange Pavilion combines robotically moulded glass reinforced concrete (GRC), 1400 pieces of router cut Australian hardwood and 250 m of stainless-steel ground rails. Studio Chris Fox designed, modelled and managed each component of the project with a computational workflow. The result is a 350 m^2 public art installation that has created a new flexible meeting place and community event space, rich with heritage meaning in this unique rail precinct.

The Pavilion references the railway switches that were once part of this area. There are five tracks on each of the three shells and they all intersect in the center of the pavilion on the ground. The tracks are used to define the aesthetic and formal quality of the pavilion but also to control the fabrication process of the elements.

The modelling process of Interchange Pavilion was hybrid. Partially manual due to the complex doubly curved shape of the pavilion and where required to have a faster and more precise outcome with computational tools. The GRC panels on the outer side were first modelled as a single element following the reference track. Based on the fabrication and transport constraint, but also on the aesthetic of the GRC, the panels have been split into smaller parts in coordination with the fabricators. The split lines also control the position of the GRC rafter. These elements are made of folded aluminum plates and they are used to connect the GRC panels to the internal structure. The GRC panels mould and support bracket placement were made only using the

Fig. 7 Interchange pavilion, Sydney, Australia, 2019–2020

Rhino® model. The inner side of the pavilion is cladded with hardwood timber battens. The battens shape and position are controlled by Grasshopper® scripts. As part of the workflow, the supporting elements of the timber (timber rafters) were placed following the main internal structure. The timber battens and rafters are all different because they follow the doubly curved shape of the pavilion.

Both the GRC panels and the timber battens are supported by the internal aluminum waffle structure. This structural solution was used to simplify the assembly process but at the same time to ensure enough stiffness for the strong wind in the area. The timber and GRC rafters also provide bracing reinforcement to the pavilion (Fig. 8).

Cut files and shop drawings were generated using the complex attributes associated to every single element during the modelling process. Elefront [8] components were important to assign, update and manage the attributes of the geometries. The timber cut files were exported directly from Rhino® as dxf file ready for production. Similarly, all the aluminum cut files have been exported from Rhino® following an attribute referenced process. The aluminum structural model was also shared with the fabricators to help them during the assembly process.

Interchange Pavilion delivery was the result of a comprehensive and collaborative approach that encompassed various stages of the project. From the early design phase, discussions and coordination were prioritized with all fabrication parties involved, allowing for valuable input to inform and optimize both the aesthetic and fabrication aspects of the pavilion. This early engagement ensured that the final outcome aligned

Fig. 8 Interchange pavilion aluminium structure, GRC and timber cladding

with the desired design quality, promoting an efficient and high-quality fabrication process.

To ensure the structural performance and fabricability of the pavilion, a continuous model exchange was established with the engineering team. This iterative process involved regular exchanges of design updates and feedback, enabling the optimization of the structure for fabrication while meeting the necessary structural requirements. This close collaboration between the design team and the engineer resulted in a structurally performing pavilion that was also well-suited for fabrication.

During the assembly phase, constant information exchange took place with the aluminum fabricators and GRC installers on-site. This proactive communication aimed to minimize mistakes and delays, promoting a smooth and accurate assembly process. By addressing any queries or concerns promptly and providing real-time information, potential errors or misinterpretations were mitigated, ensuring efficient and precise assembly.

Moreover, constant coordination with the client was maintained throughout the delivery phase. This involved regular communication, progress updates and milestone reviews to ensure that the program and the cost plan were met according to the agreed schedule.

It is noteworthy that the delivery of the Interchange Pavilion was carried out by Studio Chris Fox as the principal contractor, which entailed taking on significant management and financial risks, particularly during the Covid-19 pandemic. Delays and increased costs related to the pandemic have been managed and partially took on by SCF to minimize any major disruption during the delivery stage. Despite these challenges, SCF successfully navigated the project. This was also possible thanks to an open communication with the client that helped mitigating SCF financial exposure.

Interchange Pavilion's delivery encompassed early discussions with fabricators, continuous model exchange with the engineer, on-site information exchange, constant client coordination, all effectively managed by SCF as the principal contractor. These collaborative and adaptive strategies ensured an optimized design, streamlined fabrication and assembly for the successful delivery of Interchange Pavilion.

2.3 *Re/place Pavilion*

The fourth iteration of the Summer Place Pavilion in Albury was a structural experiment using recycled materials and incorporating circular design thinking.

Re/place Pavilion (Fig. 9) was collaboratively designed and built by Studio Chris Fox, Bollinger+Grohmann, The University of Sydney and Swinburne University of Technology. It combined the creativity and expertise of architects, artists, students, engineers, and designers. Exploring circular design thinking, the team rethought, reduced, reused, remanufactured, repurposed and recycled materials to deliver the final Pavilion design.

The core structure of the *Re/place Pavilion* was built from repurposed construction timber waste, sourced from various sites around Sydney and Melbourne. The Pavilion façade is composed of recycled timber from Albury including reclaimed elements taken from previous iterations of the Summer Place Pavilion. The facade of the timber structure was complimented by mycelium features (incorporating panels grown from fungi) that are prototypes for the design and utilization of emerging biomaterials for use in construction.

The logic of the Pavilion was based on modular construction techniques and engineered through computational design (Fig. 10). Potential material inputs were recalibrated into new structural forms leading to the resolution of the final Pavilion design. This process also helped minimize wastage in the construction and assembly phases of the project.

Circular design thinking was the key principle for Re/place Pavilion. The initial design was therefore influenced by recycled construction materials, in particular timber. Even if the material was not available and clearly defined, the first idea was to test aggregation processes starting from predefined elements with Wasp (Grasshopper Plugin). The dimensions of the timber elements have been assumed based on typical ones available on the market. Working on single elements connected to each other was challenging in terms of time and optimization for fabrication. The construction

Fig. 9 Re/place pavilion, Albury, Australia, 2022–2023

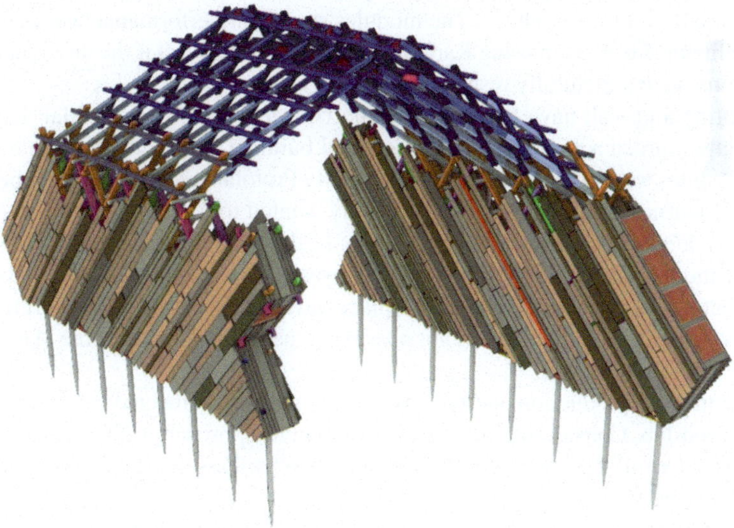

Fig. 10 Re/place pavilion final Rhino® Model

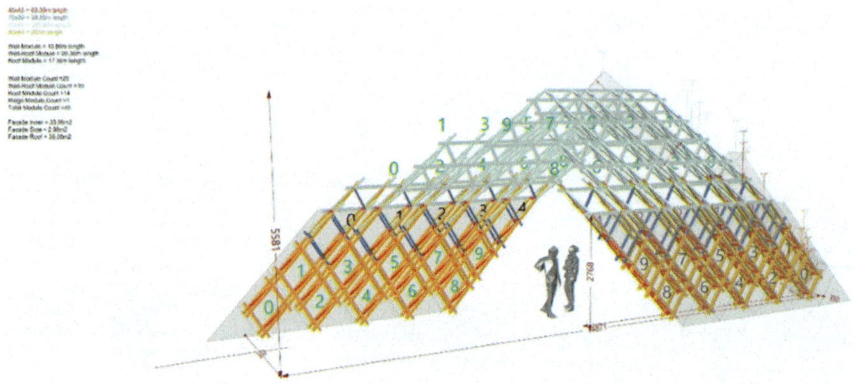

Fig. 11 Parametric model of the pavilion

documentation to for the aggregated form would have required extensive drawings set to correctly place every single component. Physarealm and Quelea (Grasshopper Plugins) have been used for design investigation. The resulting schemens were interesting, but they have been quickly abandoned mainly because of the complicated construction documentation required.

A different design approach using modules was used to simplify both the design and the construction. The final design only has four different modules: footing, wall, wall to roof and roof modules. The modules structural performance was controlled by Bollinger+Grohmann using Karamba 3D [9] to quickly test if the different design iterations were structurally responsive (Fig. 11).

Footing and wall have been built at The University of Sydney DMaF Lab with materials sourced across multiple demolished building in the city. Wall to roof and roof modules were built at Swinburne University Protolab using recycled timber from the Melbourne area. The materials originally wanted to be sourced from a building that was due to be demolished. 3D scanning of the structure has been carried out as part of the research. The scanned elements were remodelled in Rhino® to be used as inventory for the construction of Re/place Pavilion. Because of the short timeline for design and fabrication, it was not possible to use the materials of the 3D scanned source building.

The module production and pre-assembly for test was manged by the university fabrication lab. Considering the repetitive modular approach, CNC machines have been used to mill the timber beams (Fig. 12). The final assembly of all modules was made in Albury.

The Albury City Council played a significant role right from the beginning of the project, demonstrating their support for circular design thinking and the concept of reusing construction materials proposed by the design team. Their involvement and support were constant throughout the entire project. The Council did not impose any specific aesthetic requirements or directions for the design. Instead, their main focus was on reviewing the design from a safety and compliance perspective. This allowed

Fig. 12 Re/place pavilion module CNC fabrication

Studio Chris Fox and Bollinger+Grohmann design teams to have more creative freedom while ensuring that the pavilion followed the necessary safety standards.

The University of Sydney and Swinburne University had a crucial role in providing expertise and facilities supports. Researchers from both universities have been part of the team to include their research and know-how regarding the mycelium growth and use for circular design.

Throughout the entire design stage, there was a constant exchange of models and information between all the parties involved. This exchange facilitated close communication and coordination, resulting in the successful realization of the project. With every team member being across the parameters and requirements of Re/place Pavilion, the fabrication and installation stages proceeded smoothly until the project completion. Studio Chris Fox took on the responsibility of managing the assembly stage in collaboration with the installers, assuming the role of the principal contractor. This arrangement allowed for seamless coordination and efficient execution of the project.

References

1. Physarealm. https://www.food4Rhino.com/en/app/physarealm
2. Quelea—agent-based design for grasshopper. https://www.food4Rhino.com/en/app/quelea-agent-based-design-grasshopper
3. Kangaroo Physics. https://www.food4Rhino.com/en/app/kangaroo-physics
4. Grasshopper®. https://www.grasshopper3d.com
5. Octopus. https://www.food4Rhino.com/en/app/octopus
6. Geometry gym. https://geometrygym.wordpress.com
7. Rhino.Inside®. https://www.Rhino3d.com/features/Rhino-inside-revit
8. Elefront. https://www.food4Rhino.com/en/app/elefront
9. Karamba 3D. https://karamba3d.com

Chris Fox has practiced as both an artist and architect, with over 25 years' experience in the built environment, navigating the complex constraints of the public domain, to design and deliver landmark interventions of wonderment and beauty. He is passionate about transforming our cities, towns and landscapes through artful thinking and collaboration. Chris is also a Senior Lecturer in Art Processes and Architecture at the University of Sydney School of Architecture, Design and Planning.

Tommaso Pagani studied Architecture at Politecnico di Milano, where he completed the bachelor's and master's degree. During the master thesis he started to develop his passion and knowledge about computational design and digital fabrication. Tommaso is project lead and a key member in the workflow processes of the Studio Chris Fox team. He is interested in how digital fabrication and new technologies can influence the built environment.

Chapter 10
Timber Structures Through Advanced Design Tools. The Reciprocalshell Project

Christopher Robeller

Abstract The Reciprocal shell is a lightweight freeform structure, optimized for fabrication and assembly. The demonstrator investigates a new method for the digital design, fabrication and construction of timber grid shells made from simple, linear (non-curved) members. It addresses the challenge of parallel face offsetting in a triangulated surface with a catenary shaped cross-section. We solve this problem using a trivalent planar polygon topology, known from recent timber plate structure prototypes. Applying this method to a grid shell, we cross-braced each polygon using reciprocal nodes. The robotic fabrication process was optimized for a fast and simple assembly. The production process is completely carried out with only a saw blade, which allows for very fast and precise cuts. The demonstrator was built as part of an International Design and Build Workshop between Chalmers University of Technology and Augsburg University of Applied Sciences.

Keywords Digital fabrication · Shell structures · Lightweight structures · Timber construction · Robotic fabrication

1 State of the Art

Timber Grid shells are known to be very efficient wide-span structures. Famous examples include the 1983 Tacoma Dome with a 160 m span [11] or the Saldome structures (2005/2012) with a span of up to 120 m. In comparison to concrete shell structures, the timber grid shells consist of many individual members and joints, later of which are typically steel nodes that connect to the wooden members with a large amount of steel bolts or screws. These joints contribute greatly to the overall cost and complexity of the structures. Therefore, the above-mentioned timber grid shells are spherical shapes (isotropic surfaces), such as Buckminster Fuller's lattice shell structures made from steel. Typically, these structures are so called geodesic domes, which can be built from identically shaped nodes and members. Depending on the

C. Robeller (✉)
Technische Hochschule Augsburg, an der Hochschule 1, 86161 Augsburg, Germany
e-mail: christopher.robeller@hs-augsburg.de

P. Ruttico (ed.), *Coding Architecture*, Digital Innovations in Architecture, Engineering and Construction, https://doi.org/10.1007/978-3-031-47913-7_10

Fig. 1 Reciprocalshell Demonstrator, Chalmers University, Gothenburg, December 2022, built in a digital timber construction workshop with students and researchers from two Universities

overall number of segments in a geodesic dome, there might be no more than 3–5 different nodes and members, which allows for a relatively efficient "mass production" of geometrically identical parts. However, the geodesic shape will constrain the shape of the structure to spherical, constant-curvature forms, which deviate from a structurally optimal catenary cross-section (Fig. 1).

Freeform timber grid shells with other triangulation patterns (Fig. 2 left) will result in individual members and joints, which can be produced using CNC machines, but even more importantly, an exact parallel face offset is not possible [1, 2], meaning that once the height of the members is added on top of the pattern, the top of the members will not be meeting in the same point. Bannwart [3] shows that this problem can be avoided using beams where the width is greater than the height, and the offset vertex is chamfered. However, for larger structures members would need a cross-section where the height is much greater than the width, which is not possible with this method.

Freeform/catenary shaped timber grid shells can also be built with different methods, where the members are elastically bent and clamped together at intersection points. This type of structure is known from grid shells such as the Multihalle Mannheim (1978), the Weald and Downland Grid shell (2004) or very recently the Taiyuan Botanical Garden Domes (2021) with a span of up to 90 m (in the last project beams were not elastically bent, but pre-curved, CNC-produced glulam). Curved or elastically bent members allow for catenary shapes, but pre-curved members are

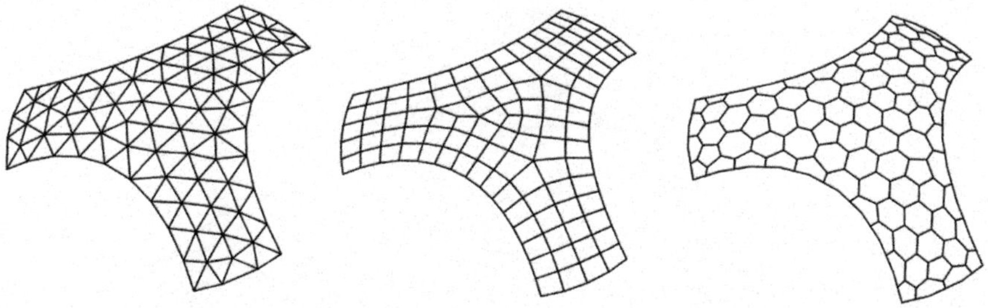

Fig. 2 Comparison of a doubly-curved, anisotropic surface with a catenary cross-section, divided into triangles (left), Quadrilaterals (middle) and a trivalent structure consisting of planar polygons with 5+ corners, allowing for a constant distance offset (right). Polygons with more than 3 corners require cross-bracing through plates, diagonals, or subdivision into triangles

considerably more expensive than linear members and elastic bending on site remains a challenging process.

2 Concept

In our Reciprocalshell project, we present a new approach to a timber grid shell, using a trivalent topology with planar facets (polygons with 5–8 corners). Such trivalent topologies (the valence at vertices is always 3) have recently been used for the design of novel timber plate structures, made from laminated veneer lumber (LVL) and cross-laminated timber (CLT). Examples are the ICD Schwäbisch Gmünd Pavillion (2014) [4], the Heilbronn BuGa Pavillion (2019) [5], the DTC Recycleshell (2019) [6] or the DTC HexBox Canopy (2019) [12]. For segmental plate shells, an exact parallel face offset [1, 2] allows for the best joining solutions. Since these structures consist of plates, cross-bracing is not needed for the polygons, even if there are more than 3 corners.

In our timber grid shell prototype (shown in Fig. 1) however, each of the polygonal faces is built as a module which consists of an outer ring of plates, which is internally cross braced through a subdivision into triangles, where the central node is reciprocal (Fig. 3, bottom). This is necessary because the surrounding corners are at various angles and cannot be connected directly with planar surfaces. Instead, we use the reciprocal node to allow for simple and efficient joints between the wooden parts. Reciprocal nodes have previously been used in various timber grid shells [7–10], however in this demonstrator we are not using it throughout the structure and only as much as possible to avoid multiple members meeting in a single point, to keep the joints as simple as possible. A known disadvantage of reciprocal nodes is, that loads are not transferred in the most direct path, which can lead to bending (and over time to creep) on members.

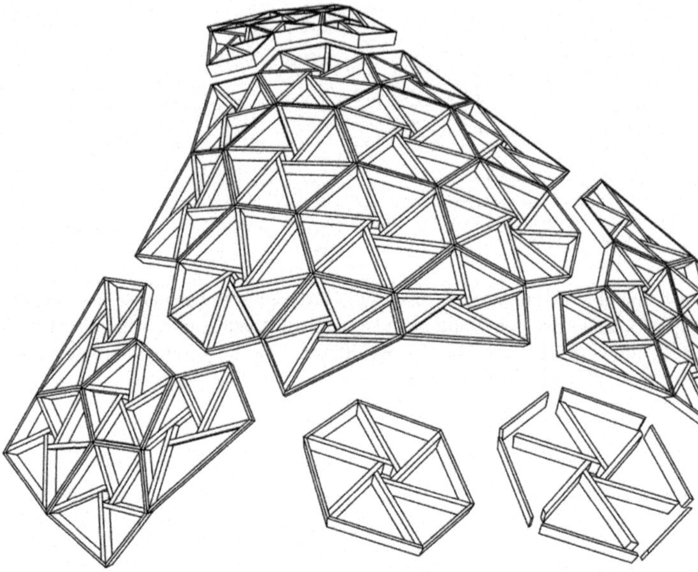

Fig. 3 Reciprocalshell Demonstrator geometry, based on a trivalent topology with planar polygonal faces. These primary polygons with 5/6 corners are cross braced through 130/40 mm members which meet in a reciprocal node at the center of the polygon

As shown in Fig. 3, the polygonal modules are joined side-by-side within the grid shell. Therefore, the outer ring of elements has the same member height than cross-bracing members, but only half of their width. Since our experimental demonstrator was relatively small (Diameter D = 7 m), we used laminated veneer lumber (LVL) strips with a width of 21 mm and a height of 130 mm for the outer ring and regular spruce wood with a width of 40 mm and a height of 130 mm for the cross-bracing members.

3 Fabrication

A major goal of the demonstrator was efficient production of the parts, using a KUKA KR150 Robot with a spindle motor attached. In our structure the polygons are planar, but the outer ring of plates in each polygon is rotated to the bisector plane between the normal vectors of the neighboring connected polygon faces. Therefore, the outer ring of parts is trapezoidal in its longitudinal cross section, meaning that it needs to be cut to individual angles not only on its short sides, but also on the two long sides.

For such many cuts, saw blades are ideal tools, since they achieve a high cutting velocity and use many blades which have a much longer cool down time before hitting the work piece again compared to milling bits. However, saw blades will

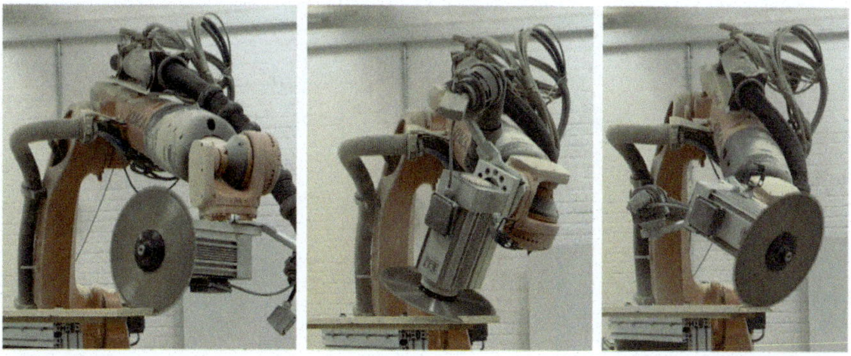

Fig. 4 Robotic fabrication setup; Left: Side cut, Middle: Mitre cut, Right: Spline cut (for the alignment of parts using plywood splines)

damage nearby parts within a tight nesting, so our workpieces needed to be clamped individually to access all four sides with the saw blade.

A custom support was built, as shown in Fig. 4. The figure also shows the three types of cuts, the longitudinal side cut, the short side miter cut, and the spline cut, which we used for the alignment and positioning of elements. For example, the position of elements at the reciprocal nodes was indicated through additional 10 mm deep and 3.5 mm wide cuts, into which we inserted 3 mm thick plywood "splines" for the precise alignment during assembly. This way we achieve a higher precision and faster assembly compared to manual measuring and positioning.

4 Assembly

As described in Sect. 3, the spline cuts and plywood splines were one of the most important strategies to allow for a precise assembly. This is very important, because any error in angles or positions would carry on to neighboring elements. In contrast to a masonry structure, our demonstrator has "no mortar between the bricks", which can help to clear tolerances that result from the production or assembly processes.

Two further strategies to deal with tolerances were the introduction of a 1 mm gap between all neighboring outer ring members, as well as large prints of the polygonal modules, which helped the assembly teams to check the correct angles between the plates (Fig. 5). As shown in this figure, the first step of the assembly was the reciprocal cross bracing members, followed by the outer ring of plywood members. The alignment was achieved through the plywood splines, then the members were glued and screwed together (screws were inserted diagonally across the wood grain). Complete polygon modules were then stored for the next step, where three legs and the center part were pre-assembled by different student teams.

Fig. 5 First step of the assembly process

The polygon-to polygon connections were achieved through two M12 steel bolts per side face, using the already existing holes that were used for the clamping during the robot cutting. Once the legs and center part were ready, the center part was lifted with a crane and temporarily supported with a stack of pallets, at a height of approximately 2 m. Then, the three "legs" were attached and connected with M12 bolts. Finally, the complete structure was once again lifted with the crane and then positioned on the three supports that were built to raise the structure 1 m above the ground.

5 Conclusion

The Reciprocalshell demonstrates a simple and cheap construction system, which proved to be easy and fast to assemble in a 2-day digital timber construction workshop with students and researchers from Chalmers University in Gothenburg and the University of applied Sciences Augsburg. Preliminary load testing showed very promising results for the structural capabilities of the demonstrator.

Trivalent, planar polygonal segmentations have recently been used for the design of timber plate structure prototypes made from laminated veneer lumber (LVL) and cross-laminated timber (CLT). Our Reciprocalshell demonstrator applies these methods to a grid shell, where each polygonal CLT module is instead built from linear wood elements with additional cross-bracing members. The cross bracing has been solved through a reciprocal node. Compared to a CLT structure, our demonstrator can save a large amount of material, since the individual members can be high (130 mm in this demonstrator) but also very slim (40/42 mm in this demonstrator). Previous grid shells with non-curved members had either used a geodesic topology, which does not allow for catenary or freeform shapes, or avoided the constant-distance-offset through members with a great width and small height, combined with cut-outs at the nodes.

Fig. 6 Team photo under the completed prototype

After the design and build workshop, the Reciprocalshell structure was load tested and will be completed with a transparent roof cover. It will be exhibited as part of the 400 year anniversary festival of the city of Gothenburg.

Fig. 7 Completed prototype

Acknowledgements The research was supported by Chalmers University of Technology and by the FNR research project No. 2220HV001X/TSI - Development of an interface for the automated structural calculation of novel, digitally prefabricated wooden structures with wood-wood connections. Project Credits: Chalmers University + Augsburg University of Applied Sciences Researchers: Amin Adelzadeh, Dr. Hamed Karimian, Karl Åhlund, Jonas Lundberg, Prof. Dr. Christopher Robeller (Chalmers Jubilee visiting Professor 2022) Students: Agathe Ducos, Adrien von der Weid, Alexandra von Bartschikowski, Anne Wicklein, Clément Braekevelt, Edith Tamm, Edona, Ege Can Yazici, Erika Perleroth, Felicia Raunås, Felix Bossenmaier, Herman Ehrnberg, Jael Kahlenberg Caso, Jakub Maliborski, Katharina Harre, Lina Svantesson, Linn Appelgren, Lluc Chia Colomer, Lorenzo Mulatero, Lukasz Gnatek, Madalin Manu, Maija Virkki, Mauritz Renz, Natasha Zwarts, Nina Tran, Nora Reis, Parjaree Manuch, Rick Persson, Rikard Murgård, Sakurako Matsuoka, Samira Sarreshtedari, Sara, Murphy Colome, Selina Ahrens, Selina Christ, Simon Ciompi, Simon Wikström, Sofia Deixelberger, Theresa Haase, William Havner, Ziming Wang. Special thanks to Chalmers University for the support of the project and to their workshop staff Tabita Nilsson, Peter Lindblom and Peter Bäckström.

References

1. Wang W, Liu Y (2010) A Note on Planar Hexagonal Meshes. In: Emiris IZ, Sottile F, Theobald T (eds) The IMA volumes in mathematics and its applications, Volume 151: nonlinear computational geometry. Springer, New York, pp 221–233
2. Hambleton D, Ross E (2015) Exact face-offsetting for polygonal meshes
3. Bannwart M (2019) The Gravitational Pavillion. Research culture in architecture: international conference on cross-disciplinary collaboration in architecture (RCA 2018). Nexus Netw J 21:425–435
4. Li JM, Knippers J (2015) Segmental timber plate shell for the landesgartenschau exhibition hall in schwäbisch gmünd—the application of finger joints in plate structures. Int J Space Struct 30(2):123–139
5. Bechert S, Sonntag D, Aldinger L, Knippers J (2021) Integrative structural design and engineering methods for segmented timber shells-BUGA Wood Pavilion. In: Structures, vol 34. Elsevier, pp 4814–4833
6. Robeller C, Von Haaren N (2020) Recycleshell: wood-only shell structures made from cross-laminated timber (CLT) production waste. J Int Assoc Shell Spat Struct 61(2):125–139
7. Schlaich M, Stavenhagen L, Krüger G (2003) Die HanseMesse in Rostock-Zollinger mit moderner Technik. Bautechnik 80(5):279–284
8. Parigi D, Kirkegaard PH (2014) Design and fabrication of free-form reciprocal structures. Nexus Netw J 16:69–87
9. Thönnissen U (2014) A form-finding instrument for reciprocal structures. Nexus Netw J 16(1):89–107
10. Robeller C (2022) Castanea Sativa Reciprocal Frame, inspired by Friedrich Zollinger, in IASS 2022 Bejing, Innovation, Sustainability and Legacy, Proceedings
11. Tacoma Dome (1983) Pacific builder and engineer
12. Robeller C, Barata EDO, Tagliaboschi EV, Schmidt-Kleespies F (2022) HexBox Canopy: a segmented timber plate shell with hardwood wedge joints. J Int Assoc Shell Spat Struct 63(4):221–231

Christopher Robeller is a Professor for Digital Design and Production at the University of Applied Sciences Augsburg, teaching and researching as part of the new and innovative program "Digital Building Master." He specializes in innovative building structures that merge architecture, structural engineering, manufacturing technology, and computer science. Christopher has previously served as a Junior Professor at TU Kaiserslautern, leading the Digital Timber Construction DTC lab, and as a Postdoc at the Swiss National Centre of Competence for Digital Fabrication NCCR dfab at ETH Zurich. Christopher holds a Doctor of Sciences degree from the Swiss Federal Institute of Technology EPFL.

Part III
Innovative Construction Industry

Chapter 11
Enhancing Efficiency in Industrial Digital Design and Fabrication of Complex Prefabricated Elements

Mirko Scaburri

Abstract This chapter discusses digitization in the construction industry and explores how the combination of advanced design tools such as parametric design and BIM with design for manufacturing and assembly systems creates a new digital chain in the AEC sector. The main focus is on describing "how digital becomes physical," which refers to the transformation of a digital model into real objects. To support the thesis, the chapter showcases several digital workflows that Styl-comp spa, a company specialized in the design and construction of special precast elements with free, complex, and custom-made shapes, implements and utilizes on a daily basis.

Keywords Digitization · Advanced design tools · Parametric design · BIM · DfMA

1 Digitalization in the Construction Sector

One of the biggest challenges in the construction industry today is definitely the digitization. Digitization today is a phenomenon that has involved and involves many industries such as automotive and aerospace. Sectors that were already highly digitized and today are aiming for an evolutionary leap, linking fully digital projects to new automated and interconnected production facilities.

The AEC (Architecture, Engineering and Construction) sector, is a very conservative and traditional, so now is having to catch up with more progressive industries.

The AEC sector's innovative approach involves transitioning from traditional drawing-based design to a more advanced model-based design methodology. This methodology incorporates three-dimensional models that not only represent geometric components but also include information attributes. This design approach

M. Scaburri (✉)
Styl-Comp S.p.A, Via Stezzano, 16, Zanica (BG), Italy
e-mail: mirko.scaburri@gmail.com

P. Ruttico (ed.), *Coding Architecture*, Digital Innovations in Architecture, Engineering and Construction, https://doi.org/10.1007/978-3-031-47913-7_11

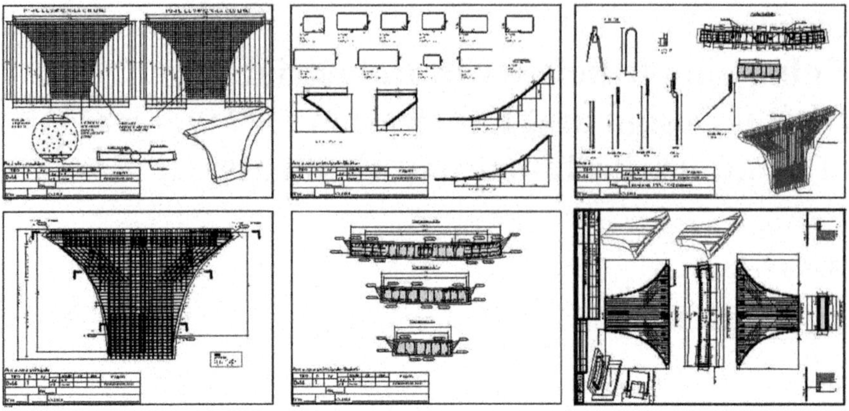

Fig. 1 Drawing-based design: shop drawings to define reinforcements and embeds of a precast element

is commonly referred to as Building Information Modeling (BIM). BIM design should not be thought of as a mere formal change from 2d to 3d, but as a new digital project infrastructure that is driving the digitization of the construction industry. The digital description of the project, through a 3d model, opens the door to new design approaches such as parametric design and generative design, but not only that, a fully digital project offers endless new possibilities even in the production and construction phases.

The program sets out a highly ambitious goal: to revolutionize an industry that primarily relies on two-dimensional design (plans, elevations, and sections) and handcrafted execution. Instead, it aims to transform the industry into a realm where everything is modeled and coded in three dimensions. In this new paradigm, all components are produced through industrial processes and then assembled on-site (Figs. 1 and 2).

1.1 Tools Drive Innovation

Parametric design and generative design in the construction world, are closely linked to one tool, Grasshopper®—a no-code programming application of the NURBS modeling software Rhinoceros®—that since its release in 2007 has gradually changed the way buildings are designed.

Parametric design is an interactive process that allows a design to be updated from a set of input parameters, such as dimensions, material type, or environmental parameters. This allows architects and engineers to make changes in real-time as the software updates the model automatically. With this approach, designers can test many more solutions than with traditional approaches and can then choose the best design option.

Fig. 2 Model based design: informative 3d model containing all the information given in the shop drawings above

Generative design, on the other hand, is an evolution of parametric design; designers always set up the design in parametric form, but computers and artificial intelligence analyzing the different variables, find hundreds of solutions that are then ranked and sorted in order to find the best one (Fig. 3).

Initially Grasshopper® was used as a formal research tool, the main objective of the tool was the so-called form-finding, succeeding in creating and managing complex, morphological forms that are difficult to manage in other design environments.

These tools have been widely used, over the past decade, making it possible to construct free-form, complex buildings, but it also laid the groundwork for the next innovative step.

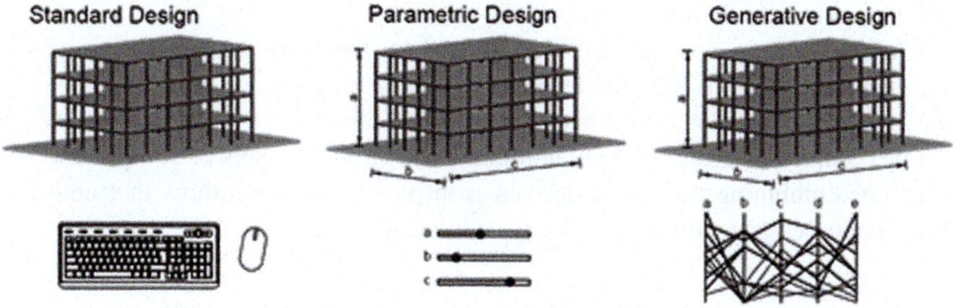

Fig. 3 Standard design versus parametric design versus generative design

Fig. 4 Palazzo Italia Expo 2015, Milan (image courtesy of Styl-Comp S.p.A)

Since 2015, a number of software houses specializing in BIM, such as Archicad® and Tekla Structures® began to create direct links with Grasshopper® [1]. These links have completely changed the idea of parametric design, greatly expanding the possibilities for use. Parametric design, relegated mainly to preliminary use and formal research only, finds new paths of development, becoming an operational tool that helps the evolution of the project, from the initial stages to the construction phase (Fig. 4).

Starting in 2020, Rhinoceros®, releases a further innovation: rhino.inside—an opensource project, which allows Rhinoceros® and Grasshopper® to work in the background within other software, such as many BIM software, such as Revit® and Tekla Structures®. This further evolution, offers the possibility of creating scripts that can be managed and used even by users who do not know how to use Grasshopper® and who generally do not use parametric design.

Thanks to these new possibilities offered by the software, parametric design becomes the engine of BIM model automation.

Conversely, the BIM model becomes a kind of container, a large database, which can be compiled indiscriminately automatically through scripts or compiled manually, thus combining the speed derived from the use of algorithms that automate repetitive processes with the flexibility of manual changes.

1.2 Design for Manufacturing and Assembly in the AEC Sector

The digitization of an entire industry should not be limited to the design phase alone. It is essential that the advantages of transitioning from "drawing-based" to "model-based" design extend to the production and assembly processes on the job site. In this regard, the AEC sector can take inspiration from technologically advanced industries like automotive and aerospace, which have successfully employed model-based design methodologies for many years.

A particularly interesting approach borrowed from the automotive sector is the Design for Manufacturing and Assembly known by the acronym DfMA. Design for Manufacture and Assembly (DfMA) is a design approach that originated in the mid-1940s when some automobile manufacturers applied it to manufacturing processes.

DfMA combines two distinct disciplines into one system: design for manufacturing (DFM) and design for assembly (DFA). Design for Manufacturing consists of simplifying the design of a product to facilitate production, with the goal of minimizing the complexity of each operation. Design for Assembly consists of design for the purpose of simplifying the assembly of a product, is concerned with reducing assembly costs and minimizing the number of operations required.

DfMA is a game-changing method of construction which involves construction being designed for manufacturing off-site in a controlled environment, before being assembled on-site.

Key benefits of DfMA include:

- Shorter construction period—Most of the construction is done off-site, reducing the construction time and the amount of manpower required on-site.
- Improved workmanship—Due to stringent quality control in prefabrication factories.
- Reduction in dis-amenities—As lesser construction work is carried out on-site, it causes less disruption to the community [2].

DfMA requires very detailed and meticulous design and therefore modeling. The goal of this design method is to create a design that is defined in every detail, so the project can be broken down into individual components that can be manufactured in the factory and then assembled on site.

One of the main characteristics of DfMA is its component driven, modularization and standardization approach. DfMA also requires planning, adapting and optimizing the design at the early stage to facilitate the fabrication of components or modules off-site and subsequently assembly on-site. Hence, the use of BIM as an object-driven tool and as an integrated collaborative environment provides potential benefits when leveraged on to drive the DfMA process [3].

To facilitate design following the DfMA (Design for Manufacturing and Assembly) approach, it is crucial to comprehend the specific production requirements for each component. For instance, within a project, some standardized components readily available in the market can be utilized. In such cases, it is sufficient to extract a comprehensive list of elements from the model, such as screws, bolts, standard profiles, plates, and so on. However, there are other components that need to be custom-made using numerically controlled machines. In these instances, it is necessary to extract machine files from the model to effectively manage the CNC machining process. Whereas for all operations carried out by people and not by machinery, a set of drawings describing the operations to be performed must be extracted from the model.

Currently, DfMA (Design for Manufacturing and Assembly) is not yet fully integrated as a process within the construction industry. Its application is primarily limited to the later stages of construction, specificaly during the construction phases involving prefabricators, metalworkers, timber frame builders, and other professionals engaged in off-site production and subsequent on-site assembly. However, during the design stages of construction, each component must be clearly defined to facilitate its actual construction.

DfMA, itself does not require the use of three-dimensional design tools, although their use simplifies implementation. Thus, commonly, many of the users of DfMA in the AEC sector do not use BIM design techniques, or at least not at a high level of detail, that can make a building constructible.

The union of three-dimensional information design (BIM) and the most advanced design and manufacturing techniques (DfMA), creates a new acronym FIM, fabrication Information Modeling.

As a BIM-based DfMA methodology, FIM provides a common interface between these two disciplines and creates a shared planning and data basis. Fabrication Information Modeling (FIM) therefore describes, analogously to BIM, a methodology for the fabrication aware integrated design, construction, and management of components in the context of an overall building using digital methods [4].

BIM and DfMA each have a different focus, require different levels of detail, and provide constraints to each other. For similar reasons, Computer Aided Design (CAD) and Computer Aided Manufacturing (CAM) have long been handled separately in the mechanical engineering industry and only recently have been more closely connected in the wake of “Industry 4.0” concepts [4].

2 From Design Idea to Construction Through Information Model

In this chapter, I will illustrate through some case histories of works carried out by Styl-Comp spa how DfMA, parametric design and BIM can be combined by creating a link between the design idea and the final product to create unique products.

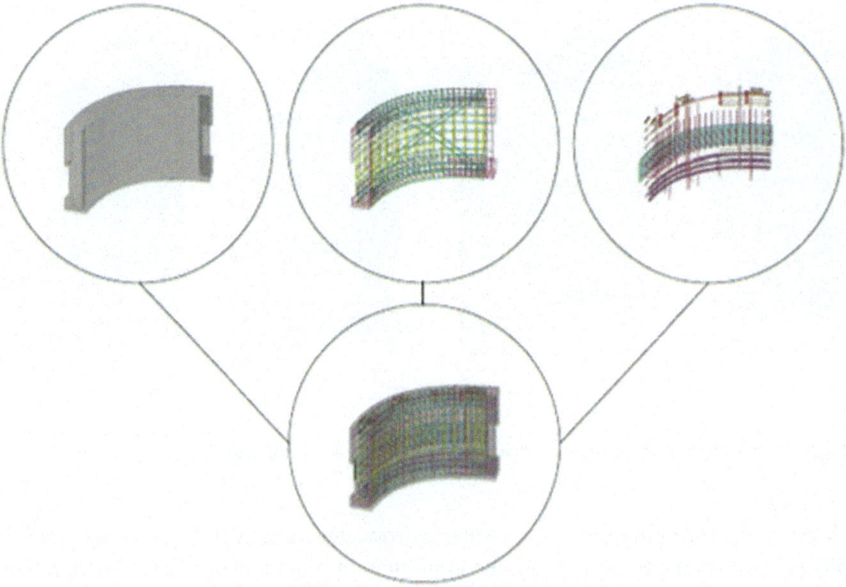

Fig. 5 Elements required for the definition of a precast element

Styl-Comp is a company that designs, builds, and installs special precast with complex, freeform, and custom shapes and seeks to take full advantage of the potential of digital and BIM model-based design [5].

The realization of prefabricated elements whether they are free-form or simpler and more linear in shape goes through the definition of three main systems: the formwork system, the reinforcement system, and the fastening system. Each of these elements is digitally defined through an informative 3d model, from which all construction information is extracted (Fig. 5).

2.1 Formwork

2.1.1 Static Formwork

A multitude of software tools are used to design formwork in Styl-Comp, a NURBS modeler (Rhinoceros®), a visual programming one (Grasshopper®), a BIM authoring tool (Tekla Structures®) and where necessary a solid modeling tool (Catia®). Depending on the geometric complexity, manual modeling or algorithm-based modeling prevails. Over time, the algorithms initially developed, have been implemented and generalized so that formwork for different shapes can be obtained, but which have the same starting algorithm (Fig. 6).

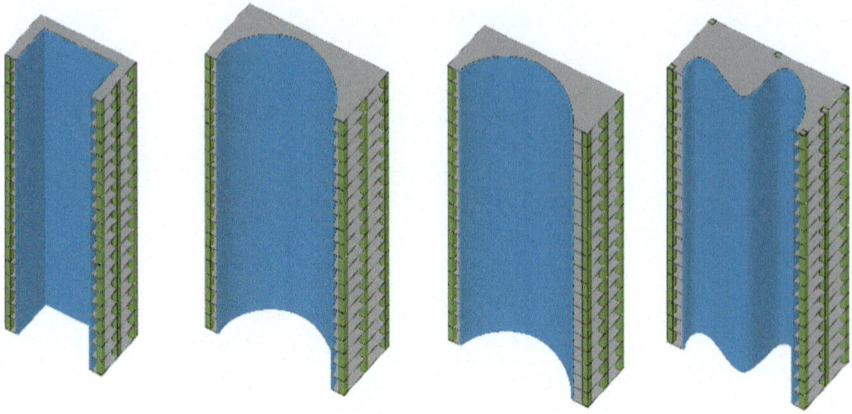

Fig. 6 Formwork of different geometries obtained by the same algorithm

The ability to obtain different shapes from the same algorithm is an inherent process of parametric design, but the combination of parametric design with a BIM modeling software, is a relatively new approach and allows the speed of parametric design to be leveraged with the flexibility of being able to introduce manual changes. In addition, the BIM authoring software used, is a tool specifically for construction design, it is then possible to manage every phase of construction directly from the BIM authoring software.

At this stage, the parametric design of the algorithms (developed in Grasshopper®) is combined with BIM modeling in Tekla Structures®, taking into account all the design criteria of the formwork and respecting the principles of DfMA. The 3d model obtained in this way, becomes a FIM (fabrication information modeling), so a model ready to be fabricated (Fig. 7).

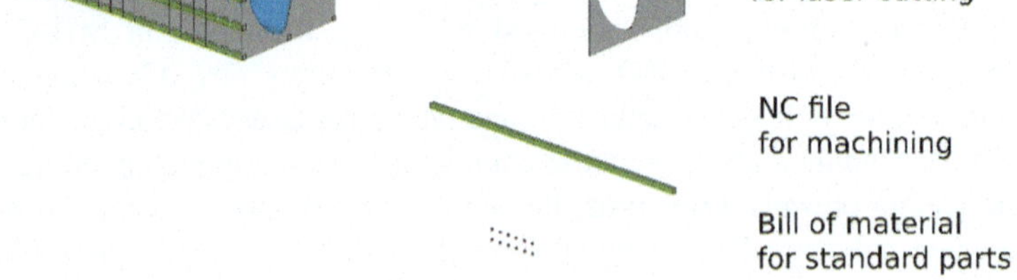

Fig. 7 Formwork components

The FIM model is a database, where all the components are stored. In order to transform these components in real things they need to be subdivided according to production or procurement methods.

Standard elements, (e.g., for screws and bolts) they can be found on the market, in this case a BOM (bill of material) is extracted from the model, which will be used to ordering components.

Steel elements, the production method for these elements is mainly based on numerically controlled machines, in this case a NC file is extracted from the FIM model. NC (Numerical Control) refers to a method where machine tool operations are controlled with a computer. The NC data controls the motion of CNC (computer numerical control) machine tools. During the manufacturing process a machine tool or machining center drills, cuts, punches, or shapes the piece of material. NC files are extracted in DSTV format (Deutscher Stahlbau-Verband). DSTV is a standard interface for geometrical description of steel structure pieces for the post-processors with numerical control. The essential aim of this interface is to be neutral, which means that with only one standard description you can manage several different NC machines. The interface standardizes the link between a CAD-program or a graphical system via a CAM file for the NC machines. The geometry of the piece is introduced completely neutrally, and after knowing the parameters of the NC machine, the post-processor is able to translate this neutral language to the NC machine language [6] (Fig. 8).

```
P/6
1
Steel_Undefined
21
PL10
B
  1279.07
   532.14
    10.00
    10.00
    10.00
     0.00
   78.500
    2.115
    0.000
    0.000
    0.000
    0.000
```

```
AK
  v       0.00u       0.00       0.00       0.00       0.00       0.00       0.00
       1279.07        0.00       0.00       0.00       0.00       0.00       0.00
       1279.07      532.14       0.00       0.00       0.00       0.00       0.00
       1178.99      532.14       0.00       0.00       0.00       0.00       0.00
```

Fig. 8 NC files for steel plate production

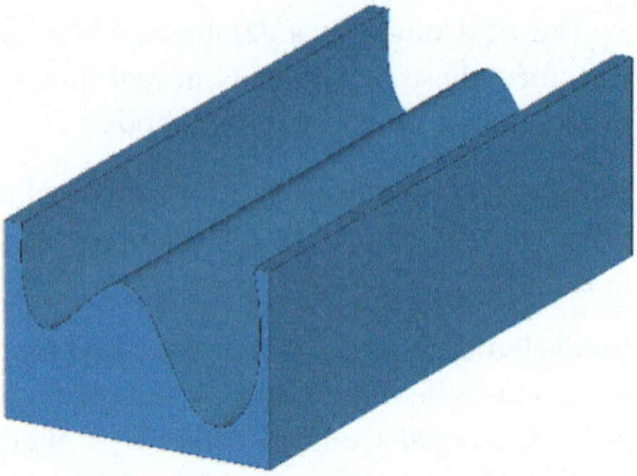

Fig. 9 STP shape for creating GCODE for milling

Formwork surface, if the precast elements have a particularly complex shapes, the formwork may require numerically controlled milled molds. In these cases a sub-model is extracted from the FIM model and exported in STP format, the STP file need be processed in an external software (like Sprutcam® or directly in Grasshopper®) for the generation of the machine file (gcode) intended for the numerically controlled milling machine (Fig. 9).

At the end, all the ordered, and machined components need to be assembled in the factory. In this phase, drawings are strictly required, the assembly phase is made by human operators, who require drawings with all the instructions included.

To create this kind of assembly drawings we still use the FIM model, in this way we can create detailed assembly drawing with cross reference information, so the name of each sub-component is shown in the assembly drawing (Fig. 10).

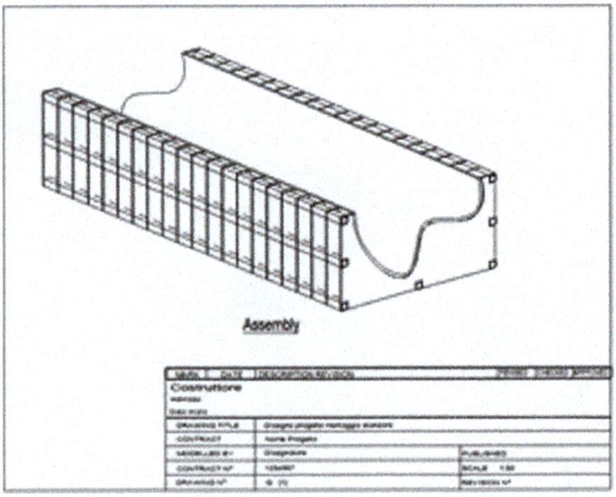

Fig. 10 Assembly drawing for the formwork

The digital workflow described above takes shape in each work carried out by Styl-Comp. As an example, we report here a work called "Jardin Exotique" carried out in the principality of Monaco. The project involves the creation of a luxury residential complex with a 15-story underground parking garage and a rooftop garden bridging the existing road. The project was designed by Fabrice Notari and Rudy Ricciotti Architects (Fig. 14). Styl-Comp was involved to design, produce and install the precast structures of the bridge roof garden. The project consisted of 9 free-form pillars, 50 flat elements type predalles, and 50 curved elements called coques. The pillars are characterized by a free form, mushroom or tree shape with a base width of 0.9 m and a top section of about 2.4 m. In order to comprehensively demonstrate the digital workflow employed, particular emphasis is placed on the column in this report.

As seen previously in Fig. 5, the initial crucial step in producing the final precast product (Fig. 13) involves creating a FIM model of the formwork (Fig. 11). This is accomplished using the algorithms described earlier. Once the model is prepared, it is divided into sub-components for manufacturing. After all the components are ready, they are assembled as depicted in Fig. 12.

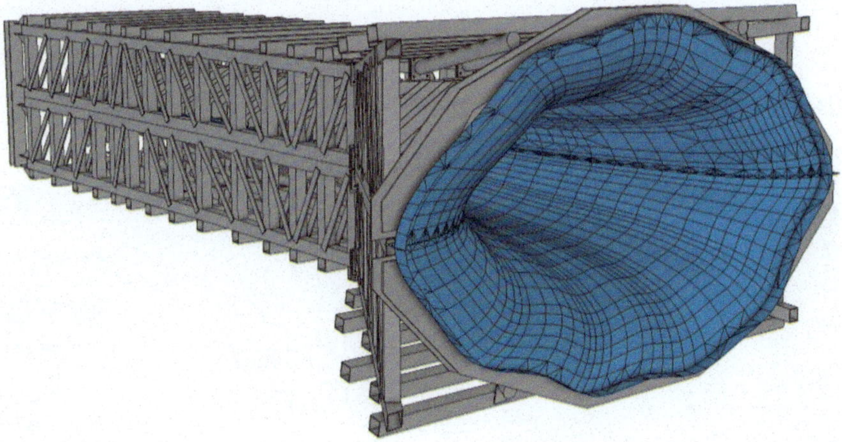

Fig. 11 FIM model of the column's formwork

Fig. 12 Formwork structure (image courtesy of Styl-Comp S.p.A)

Fig. 13 Finished column (image courtesy of Styl-Comp S.p.A)

Fig. 14 Photograph of the complete work (image courtesy of Styl-Comp S.p.A, photographer Stefano Anzini)

2.1.2 Adaptive Formwork

Given the strong technological propensity of Styl-Comp, a new technology was introduced in 2020 that fits perfectly into the workflow described above. As we saw in Sect. 2.1.1 starting with the shapes required by design, a series of digital workflows were created in Styl-Comp to automatically or semi-automatically create formwork. However, the ad hoc creation of formwork, used only once, has a significant impact on production time and cost. In order to get around this problem, Styl-Comp thanks to Adapa S/A has created 3 adaptive formworks, which allows, as opposed to static formwork to be reused an infinite number of times, ensuring savings not only in time and cost, but also in environmental impact.

However, the digital process follows a similar procedure to the design of static formwork; once the shape is defined, the surface is processed within Rhinoceros and Grasshopper® where the XML file is generated and then read by the machine. In this way the digital information instead of being used to produce static and disposable formwork, is used for the programming of dynamic formwork that depending on the input derived from the 3d model adapts, recreating in reality the shape displayed in the virtual model (Figs. 15, 16 and 17).

Fig. 15 ADAPA double curvature adaptive formwork (image courtesy of ADAPA S/A)

Fig. 16 Precast curved piece (image courtesy of Styl-Comp S.p.A)

Fig. 17 Single-curved adaptive formwork with vertical tilting system (image courtesy of Styl-Comp S.p.A)

2.2 *Steel Reinforcement*

When designing special, free-form precast elements, one of the most complex parts is the design and implementation of the reinforcement system. An element, even one that is not too complex in shape, such as a curved precast piece with inclined faces, can involve hundreds of reinforcing bars with different lengths and shapes. This complexity must be able to be managed and solved at the design level, while also taking into account production needs, such as the steel bending machine and assembly difficulties (Fig. 18).

Also, for the reinforcements, in Styl-Comp we have set up a series of scripts for the automatic creation of reinforcement, to facilitate the design part. Starting from the geometry of the precast element, a series of scripts, generate stirrups and reinforcement, elements that are processed through Grasshopper® and then are managed within Tekla Structures® (Fig. 19).

Parametric modeling of reinforcement not only offers faster shop drawing preparation than manual modeling, but also increases its quality, ensuring the correctness and complete geometric congruence of each element.

Once the parametric design phase is finished, the modeled rebar elements are sent and managed within the BIM authoring software, where through additional macros and custom scripts, the cage assembly drawings and rebar list are extracted.

Fig. 18 Column reinforcement of Fig. 13 (image courtesy of Styl-Comp S.p.A)

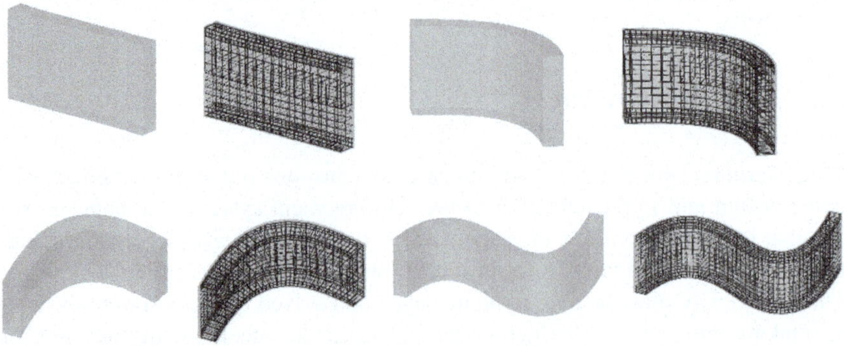

Fig. 19 Precast reinforcement with different geometries obtained from the same script (image courtesy of Nanoo)

As explained for formwork, in the previous chapter, an extremely important aspect once detailed information modeling has been obtained is to link this digital information to production systems, that means, how the information contained in the 3d model can be transformed into real objects.

In this instance, the reinforcing bars were extracted from the model and transmitted to the steel bending factory in the BVBS (Bundesvereinigung Bausoftware) format. For metal beams, a similar procedure is followed, utilizing a text file with ASCII code and a ".abs" extension. The BVBS format was collaboratively developed by bending machine factories, construction software companies, reinforcement bending works, steel producers, and academic institutions. Its purpose is to facilitate the exchange of information between rebar detailing software and CNC controlled bending processes.

```
BF2D@HjC 202100@rA-32@i0@pA-32.15@l1690@n2@e0.83@d14@gB450C@s0@a0@t0@Gl1689@w0@C80@
BF2D@HjC 202100@rA-32@i0@pA-32.14@l1300@n20@e1.57@d14@gB450C@s56@a0@t0@Gl1299@w0@C81@
```

Fig. 20 Example of BVBS strings

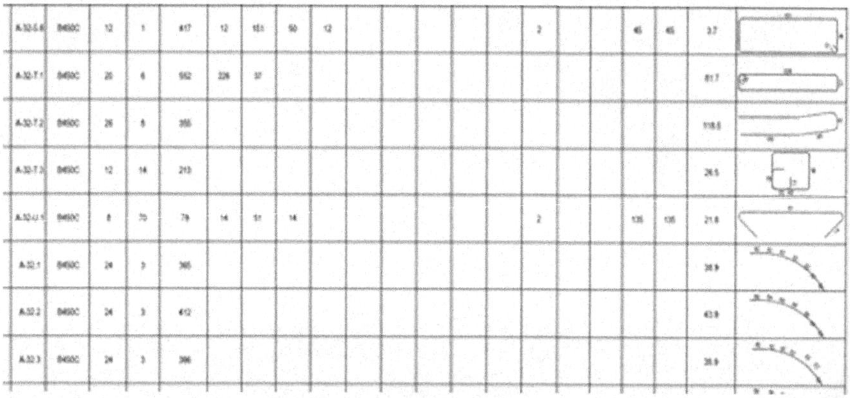

Fig. 21 List of reinforcing rebars extracted from the model

The BVBS specification features a data structure from the designer's perspective regardless of the fabrication machine which will be used and may also be understandable without authoring CAD/BIM tool. The reinforcement data are exchanged through a ASCII encoded text file and, as other CAD-CAM formats. These files are composed by a data string divided into blocks and preceded by a recognition code used to set the shape type group [7] (Fig. 20).

Also in this case, as per formwork the BIM model becomes a FIM model, so it is possible to extract directly form the model a BVBS list (Fig. 19) that can be processed directly by the bar bending machines. Contextually, a rebar list can be extracted from the model, (automatically), this file is more human-readable than the BVBS file, and it was used as an order list and checklist (Fig. 21).

By employing this process, it becomes achievable to completely eliminate the need for manual rebar extraction and dimensioning operations, which are traditionally conducted to provide steelworkers with the necessary information for rebar supply and bending. Through this fully digital workflow, the design, production, and bending of reinforcing bars can be carried out efficiently and without any loss of information.

However, the construction of the reinforcing cage is still done manually, so once the bars are received already bent, as designed, they must be assembled at the plant. In order to do this, operators need paper assembly diagrams, which are also extracted directly from the FIM model (Figs. 22 and 23).

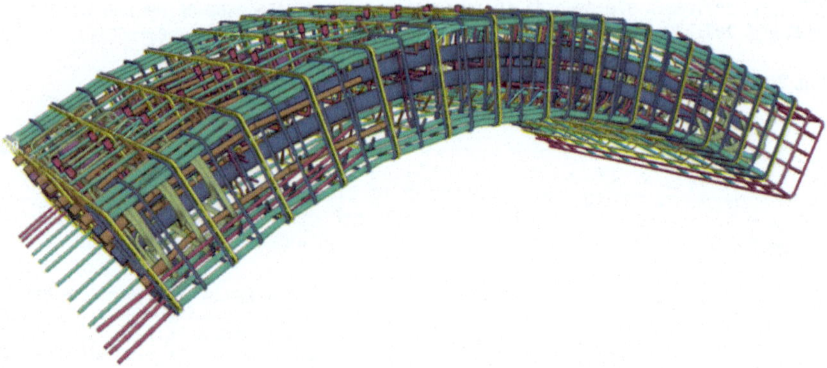

Fig. 22 FIM model of the reinforcements (image courtesy of Styl-Comp S.p.A)

Fig. 23 Steel reinforcement cage during the assembly phase (image courtesy of Styl-Comp S.p.A)

3 Conclusion

Building a digital chain is a major challenge for the construction industry today; the increasingly common use of information modeling (BIM) is a key link in this new chain. However, as we saw in the chapter, BIM modeling is not enough; it is critical to manage the transition from information model (BIM) to a manufacturing

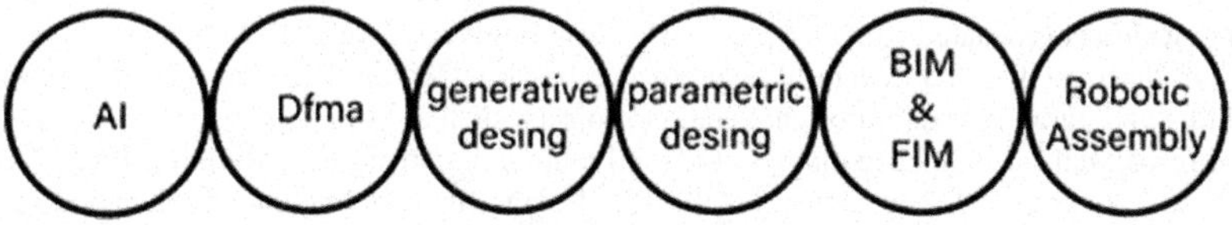

Fig. 24 Digital chain (image courtesy of Nanoo)

information model (FIM) [8] considering all manufacturing and assembly needs. In this way, the large amount of model data can be used not only with the descriptive and design purpose but also with the manufacturing purpose.

To implement this shift, it is necessary for systems such as design for manufacturing and assembly to make a quantum leap from the construction phase, in which they are used today, to the preliminary design stages.

This anticipation of detailed design can be made possible precisely through the joint use of BIM tools and parametric design, tools that enable high levels of detail to be achieved, relatively quickly.

The digitization process in the AEC sector today is biased, and some links in the digital chain are not yet fully connected and developed. The extensive use of advanced digital tools and DfMA design in the preliminary stages will bring about a profound change in the AEC sector.

The previously discussed digital chain, encompassing DfMA, generative design, parametric design, and FIM, will be further enhanced by incorporating AI into the initial stages. This involves leveraging AI for tasks such as scripting design parameters. Additionally, the integration of robots and automated systems will not be limited to certain production stages but will extend to the assembly phases as well [9] (Fig. 24).

References

1. Wojslaw K (2022) Grasshopper in BIM. BIMcorner
2. DfMA guide line in Singapore, https://www1.bca.gov.sg/buildsg/productivity/design-for-manufacturing-and-assembly-dfma
3. BIM for DfMA (Design for Manufacturing and Assembly) (2016) Essential Guide Building and Construction Authority and Bryden Wood, Singapore
4. Slepicka M, Vilgertshofer S, Borrmann A (2021) Fabrication information modeling. In: International symposium on automation and robotics in construction (ISARC 2021)
5. For further information refer to Styl-comp.it
6. Standard description for steel structure pieces for the numerical controls (1998) DSTV Commissions
7. Maciel AR, Corrêa FR (2016) Interoperability with IFC in the automated rebar fabrication. In: 33rd International symposium on automation and robotics in construction (ISARC 2016)
8. For further information refer to Nanoo.cc
9. Makris S, Mourtzis D, Chryssolouris G (2014) Computer aided manufacturing. CIRP Encyclopedia of Production Engineering, pp 254–266

Mirko Scaburri graduated with an M.Sc. in Building Engineering at UNIBG in 2014. In the same year, he joined Redesco, a structural firm practice based in Milan, where he was involved in 3D modeling and parameterization of challenging projects, such as skyscrapers, museums, shopping malls, residential, and office buildings. Currently, he is employed as a Building Engineer/BIM expert at Styl-comp spa, where he is involved in design, 3D modeling, and the creation of parametric scripts for the construction of special pre-fabricated and architectural concrete facades. In 2020, he launched a personal project called Nanoo, with the idea of sharing his experience and improving automation in the AEC sector.

Chapter 12
Transforming the Construction Industry Through Wire Arc Additive Manufacturing

Christopher Borg Costanzi, Benedikt Waldschmitt, Ulrich Knaack, and Jörg Lange

Abstract This chapter presents six research projects conducted in collaboration with academic and industrial partners at TU Darmstadt, investigating the use of Wire Arc Additive Manufacturing (WAAM) in Architecture, Engineering, and Construction (AEC). The case studies demonstrate the versatility of WAAM applications, including function integration, mass customization, onsite fabrication, and post-processing. The studies cover: (i) Hybrid reinforcement of thin sheet metal for function integration and mass customization. (ii) WAAM-manufactured lattice structures for structural design. (iii) Individualized connectors for glass facades showcasing mass customization challenges. (iv) Onsite fabrication of a steel printed bridge, identifying welding challenges. (v/vi) Post-processing techniques using milling to create complex shapes and textures, along with Rudimentary Digital twins for process control. The chapter highlights WAAM's potential to transform AEC and identifies key challenges, offering valuable insights and suggesting future research directions.

Keywords WAAM · Mass-customization · Structural design · On-site fabrication

C. B. Costanzi (✉)
Institute of Structural Mechanics and Design, Darmstadt Technical University, Darmstadt, Germany
e-mail: Christopher.Costanzi@Brunel.net

B. Waldschmitt · J. Lange
Institute for Steel Construction and Materials Mechanics, Darmstadt, Germany
e-mail: waldschmitt@stahlbau.tu-darmstadt.de

J. Lange
e-mail: Lange@stahlbau.tu-darmstadt.de

U. Knaack
Institute for Structural Mechanics and Design, Darmstadt, Germany
e-mail: Knaack@Ismd.tu-darmtstadt.de

P. Ruttico (ed.), *Coding Architecture*, Digital Innovations in Architecture, Engineering and Construction, https://doi.org/10.1007/978-3-031-47913-7_12

1 Introduction

Wire Arc Additive Manufacturing (WAAM) is a rapidly developing technology which, in recent years, has shown significant potential for revolutionizing the manufacturing of metallic components. Based on a process in which an electric arc is used to melt a wire feedstock, WAAM offers the similar benefits found in other Additive Manufacturing (AM) techniques. This includes the potential for cost-effective and flexible alternatives to traditional manufacturing methods which must be quick to adapt to an ever-increasingly digitized world of fabrication. In Architecture and Engineering Construction Industry (AEC), this technology has already been used to fabricate geometrically-complex structural elements which are either unfeasible or impossible to produce using conventional fabrication techniques [1, 2], Material-Efficient Structures [3] as well as large-scale building structures [4].

As with any new technology, there are numerous challenges which must be overcome before it is introduced to the public domain. This chapter outlines a number of WAAM research and industrial projects carried out at Technical University of Darmstadt and positions them in the context of a rapidly-evolving AEC industry. It illustrates both the potentials of using WAAM to rethink our structures and develop new fabrication processes, while shining a light on challenges which must also be overcome.

1.1 Background to the Technology

WAAM is an AM technology which uses a Gas Metal Arc Welding (GMAW) to precisely deposit molten weld material. The WAAM process involves the feeding of a wire electrode through a welding torch. An high-temperature electric arc is used to melt the tip of the electrode, causing it to form a molten droplet. The arc is formed by creating an electrical circuit between the welding wire and a metallic base material; when the welding torch is brought close to the base material, electrical currency flows from the welding wire to the base material, resulting in a high-temperature arc which melts the tip of the wire electrode, causing the formation of a molten droplet. The molten droplet is precisely deposited in layers, typically using Computer Numeric Control (CNC) machinery or robotic arms. Further to the arc, a shielding gas is used to protect the molten material from the surrounding atmosphere, which can cause oxidation and other defects in welded structures.

1.2 WAAM in Industry

The ability to fabricate metallic structures at production rates as high as 10 kg/hr [5], utilizing alloys including steel [6], aluminum [7], titanium [8], as well as material combinations [9], has made WAAM an attractive fabrication process in industries including, but not limited to, Aerospace, Oil and Gas, and Research and Development sectors [10]. In 2015, aerospace engineers at Stelia used WAAM to reinforce an aluminum sheet fuselage, leveraging topology optimization to efficiently generate structurally-optimized reinforcement patterns which would be otherwise unfeasible working with conventional fabrication methods [11]. In 2022, researchers from Nova School of Science and Technology, collaborated with the University of Waterloo and the Max-Planck Institute in Germany to fabricate functionally-graded materials in steel and copper, using an adapted Twin-Wire Arc Additive Manufacturing (T-WAAM) process [9]. This technique allowed the team to manufacture welded structures with varying material composition and properties, opening up a whole new potential for highly customized and performance-enhancing materials.

1.3 WAAM in Architecture, Engineering and Construction (AEC)

In the Architecture, Engineering and Construction (AEC) industry, the ability to fabricate geometrically intricate structures in a variety of materials, scales and production rates has also made WAAM an increasingly appealing tool. Its use is partially enabled through the increased uptake of computational design tools and is therefore not surprising that the implementation of WAAM in the AEC industry comes hand-in-hand with a transition towards advanced digital design and fabrication processes and technologies, which define the fourth industrial revolution we are currently experiencing.

One of the earliest adopters of the technology in AEC comes from the Dutch group MX3D who, in 2013 presented plans for the world's first bridge structure which was to be produced in WAAM [4]. The 12 m-long bridge was installed on the canals of Amsterdam in 2021 and, to date, still remains the largest WAAM-produced part in the AEC sector. Apart from producing large-scale structures, WAAM has also been used for detailing and joining standard steel elements. The researchers at ETH Zurich [12], used a robotic spot-welding process to join standard CHS- sections in a geometrically complex configuration in three-dimensional space. The researchers at Bologna University, together with MX3D [13], also used a spot-welding process to create intricate and structurally-optimized lattice column structures—showcasing the potentials for light-weight construction within the AEC. Also contributing to the theme of structurally-optimized components were the researchers at TU Darmstadt [14], LASIMM [15] and MX3D [16].

The relative speed in which structures can be produced in WAAM has also made it an ideal candidate for producing small parts with slight geometric variations. Feucht et al. shows how WAAM can be used to rapidly produce simple, yet customizable, IPE-Beam connections [17] and Stiffeners [18]. The researchers of TU Braunschweig further built upon this by introducing customization through structural performance

[19]. Nonetheless, the majority of the applications of WAAM within the AEC still remain within the context of research and academia, and is still yet to have the same uptake as other construction materials [20].

2 Methodology

The following chapter will present WAAM research projects carried out by the TEchnical University Darmstadt within the context of AEC. The projects are first presented in isolation, structured as separate sub-headings, and later compared with one another. The case studies are intended to highlight both research areas of innovation, as well as challenges which must be faced before being fully adopted within the AEC industry. The projects were primarily produced using a 6-axis robotic welding device located in the WAAM Laboratory at the Technical University of Darmstadt, which is described in the following section. The 'Spider Glass Connectors' project, however, was manufactured using a 3-axis WAAM setup by the researchers at the Technical University Ilmenau.

2.1 Wire Arc Additive Manufacturing Setup

Although the research projects all share similar WAAM setups, the primary distinction between projects lies in (i) the parent material used, (ii) the welding parameters and, in the case of the steel-printed bridge, (iii) the surrounding laboratory conditions.

Figure 1 illustrates the fundamental setup used at the laboratory of TU Darmstadt, summarized as: (a) Fronius CMT 4000 Advanced welding device, which is affixed to a (b) COMAU Smart NM 6-Axis Robotic Arm; (c) A Weko wire electrode of diameter 1.0–1.2 mm; (d) Ferroline C6X1 and C18 shielding gas; (e) COMAU C4G controller and (f) Welding table containing workpieces.

3 Presentation of Case Studies

Table 1 below presents an overview of the case study projects. They are classified by: (i) the nature of the parent material—that is, whether the objects are printed onto a sacrificial baseplate or integrated with other components, (ii) Scale and (iii) Context—Facade, Structure or Infrastructure. Keywords are also used to further contextualize each project. Mass customization, for example, refers to projects which show the potential for large-batch fabrication of bespoke components. Parametric Robotic Programming (PRP), on the other hand, refers to a computational design process that uses mathematical functions to generate robotic toolpaths, which is elaborated on in the case studies.

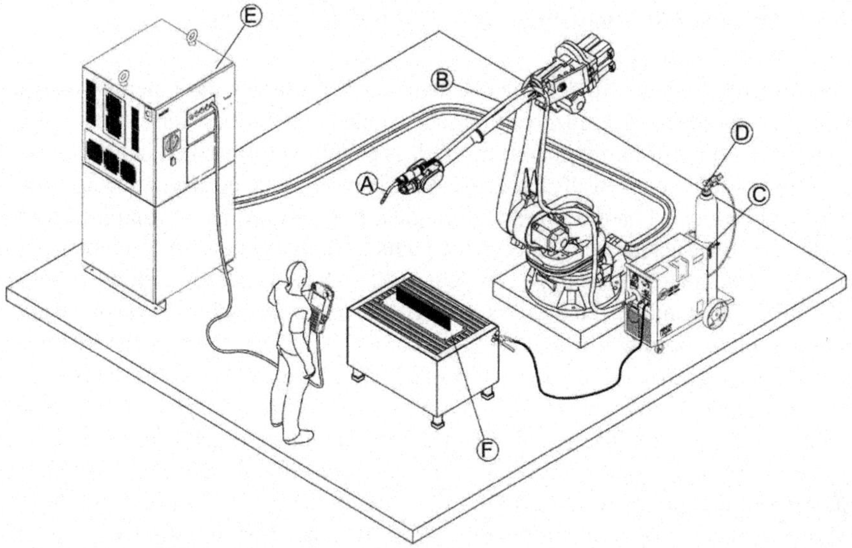

Fig. 1 Schematic of WAAM setup at TU Darmstadt consisting of **a** welding device, **b** 6-axis robotic arm, **c** wire-feed, **d** gas-feed, **e** C4G Controller and **f** welding table

Table 1 Summary of case studies presented

Case study	Parent material	Scale	Project keywords
Spider glass connectors (TU Ilmenau)	Steel facade connection plates	1:1, Facade components	1:1 Facade components, mass customization, function integration
Lattice structures	Sacrificial parent material	1:20, Structural components	Lattice structure, form finding, geometry complexity
Reinforced thin sheet metal	Integrated thin sheet metal	1:1 Facade components	1:1 Facade components, mass customization, function integration
Branching column structure	Integrated structural connection	1:1 Structural component	1:1 Structural component, Post Processing
Column structure	Sacrificial parent material	1:10 Structural component	Scaled structure, parametric robotic programming, adaptive process control
Steel bridge	35 mm Steel foundation baseplates	1:1 Infrastructure	Onsite fabrication, parametric robotic programming

3.1 Mass Customization: Spider Glass Connectors

The research project "Spider Glass Connectors" illustrates the potential of using WAAM to produce customizable interface connections for free-form, glass shingle facades. Twenty seven rhombus-shaped glass panels were arranged in a rectangular grid across a concrete wall. Each glass panel was mounted using standard cleat connections located on the edges of each glass panel. As with any ventilated facade system, a substructure is required which acts as a connection interface between a primary structure (in this case, a concrete wall) and the facade mounting connection (the cleats). In standardized facade systems, this is generally achieved using repeated, mass-produced profiles. In the case of the *Spider Glass Connectors* project, geometric complexity was introduced into the layout by (i) varying the offset of the glass panels from the concrete substructure and (ii) varying the rotation of the glass panels relative to the concrete wall. The result was a facade layout in which every single cleat connection had a unique relative offset and orientation to the concrete wall and, as such, required a uniquely fabricated sub structure for every connection point (Fig. 2). This presented an opportunity to adopt WAAM to fabricate the sub structure, as it had already been established as a flexible fabrication process which allowed for the relatively quick buildup of metal structures.

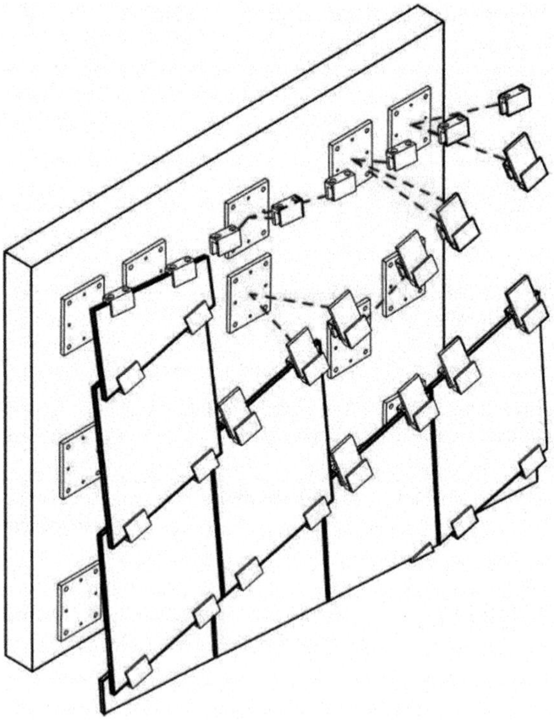

Fig. 2 Schematic layout of a section of the mockup showing a concrete wall with steel base plates affixed to it. Steel connection cleats are positioned at the corner of each glass panel, their offset and orientation to the concrete wall are represented by a dotted line

Adopting a fabrication-informed design process

Although WAAM is flexible in the scale and nature of printable geometry, it is not to say that it does not have any limitations. In the *Spider Glass Connectors* project, a 3-Axis welding setup was used. This imposed a geometric restriction in that the maximum angle in which successive layers could be deposited without the need for support structure (overhang angle) was 25° (Fig. 3a) Moreover, the welding parameters used allowed for a constant wall thickness of around 5 mm. Because the project involved facade construction, which typically requires a large number of connections between components, it was important to ensure that the design's geometry could be generated computationally with a high degree of adaptability. This meant that the design process had to account for geometry generation, structural analysis, and communication with fabrication as seamlessly as possible. Therefore, some processes such as topology optimization, which are commonly adopted in additive manufacturing, and can generate efficient structures but require significant computational effort, were not feasible due to the high number of components. Instead, the geometry was described using only curves that represented a piped geometry. The curves were generated using a combination of Grasshopper 3D to impose geometric restrictions, Karamba 3D to perform preliminary calculations, and Galapagos to generate best-fit solutions based on the requirements from Grasshopper and Karamba. While the results were not the most structurally efficient solutions, they were fabrication-aware, allowing for all connections to be generated nearly instantly while taking into account structural and fabrication requirements.

In order to minimize the amount of excess material created during the start of a Wire Arc Additive Manufacturing (WAAM) process, printing was done on specially designed base plates. These plates were customized to have the same clamping details used during the welding process, as well as mounting points on the concrete wall

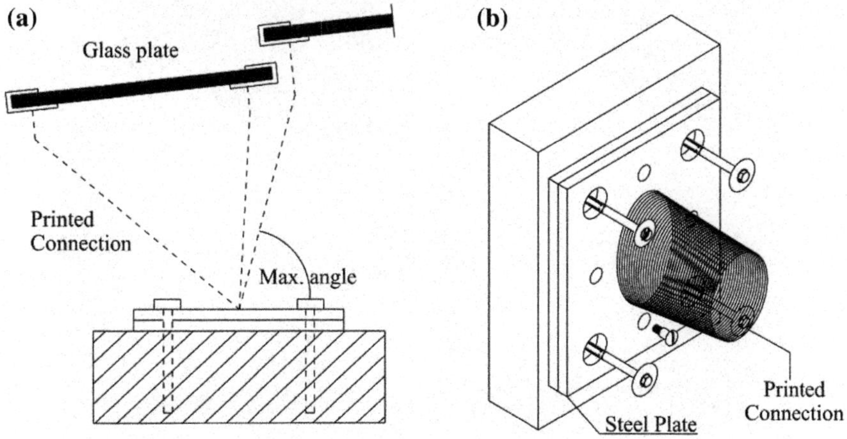

Fig. 3 **a** Schematic of the maximum overhang angle between the concrete wall/base plate and the cleats, and **b** a schematic showing how the printed connections are fixed to the concrete wall

structure, which contained pre-positioned connection plates (Fig. 3b). This approach helped to further reduce the amount of waste generated during the process.

Figure 4a illustrates the 3-Axis welding setup at TU Ilmenau, with a typical 4-point connection detail. To address the problem of relatively low-resolution inherent with the WAAM process, the printed connection branches were intentionally made longer and post-processed to the correct dimensions using robotic milling. The printed connectors were then manually screwed to connection plates which were pre-installed onto the concrete wall (Fig. 4b).

While additive manufacturing (AM) may never be as fast as conventional mass-production techniques, the *Spiderglass* project demonstrated the potential of using

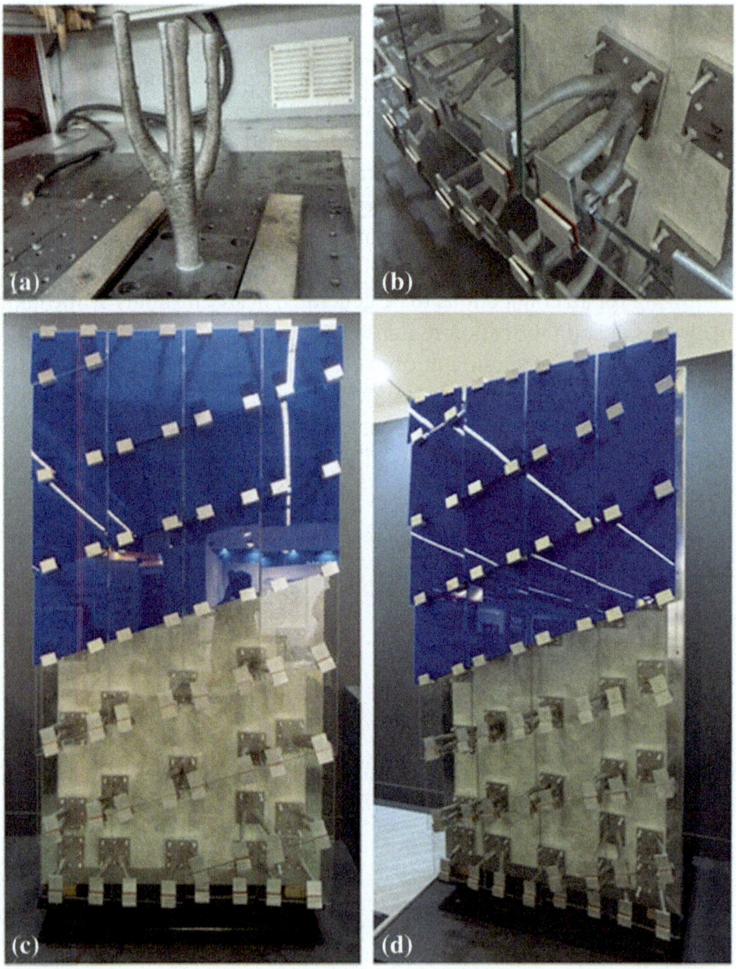

Fig. 4 **a** Welding setup at TU Ilmenau **b** Spider glass connectors were screwed to pre-existing steel plates installed on the concrete wall structure **c**, **d** The final mockup with glass plates

WAAM for mass customization. It also emphasized the importance of integrated form-finding and fabrication strategies for mass customization, as well as the potential use of post-processing to address the low resolution of WAAM.

3.2 Lightweight Structures: 3D Printed Lattice Structures

A lattice structure is a repeating pattern of interconnected struts and nodes, forming three-dimensional cells. Over the recent years, AM of lattice structures has gained popularity in various industries due to the ability to produce lightweight structures with excellent structural performance. The AEC sector has also been adopting this technology, utilizing additively-manufactured lattices in diverse applications, in materials including concrete [20, 21] and plastics [22, 23]. In metals, Laghi et al. [13] explored the use of WAAM for producing large-scale lattice structures for the AEC industry. Building upon this research, TU Darmstadt conducted a study in two parts. First, preliminary welding parameter studies were carried out to produce vertical and inclined struts of varying thicknesses and orientations, establishing a range of reasonable constraints for fabrication (Fig. 5). This information was used to set limitations for strut sizes and inclinations in a digital form-finding process.

In the form-finding process, an initial surface model representing a circular column under uniaxial load generated Euler buckling curves and moment values that were used to distribute materials in a second *initial density lattice model*. The network density was then adjusted to rapidly generate numerous solutions evaluated for maximum deflection and utilization using Karamba 3D and Galapagos, while considering the negotiation between member inclination angles for resisting bending and shear forces (Fig. 6).

Six lattice model instances were then fabricated at the TU Darmstadt robotic WAAM lab. The robot motion instructions were developed using Grasshopper 3D, using the following welding logic: (i) The lattice line model into individual line segments; (ii) Each segment was divided into equally-distributed coordinates along

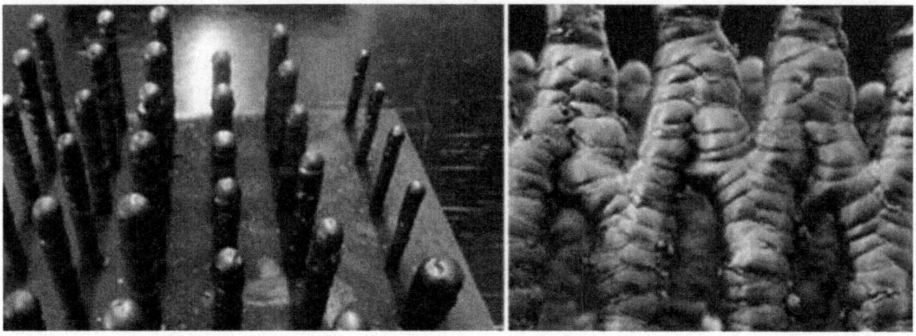

Fig. 5 Welding parameter testing for varying strut thickness were complimented with the study of maximum welding overhang using different welding processes

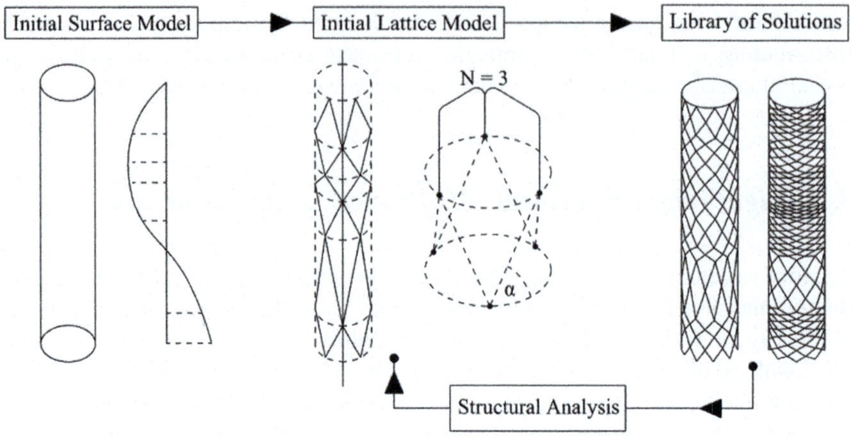

Fig. 6 An initial surface model is subjected to a uniaxial load inside Karamba 3D. The corresponding force diagrams are used to distribute the density of an initial lattice model

its length, representing the travel and welding positions; (iii) Points were sorted based on their z-coordinate values, maintaining a bottom-to-top welding sequence; (iv) for each layer-group, the coordinates were radially to avoid neighboring points. This process is summarized below in Fig. 7.

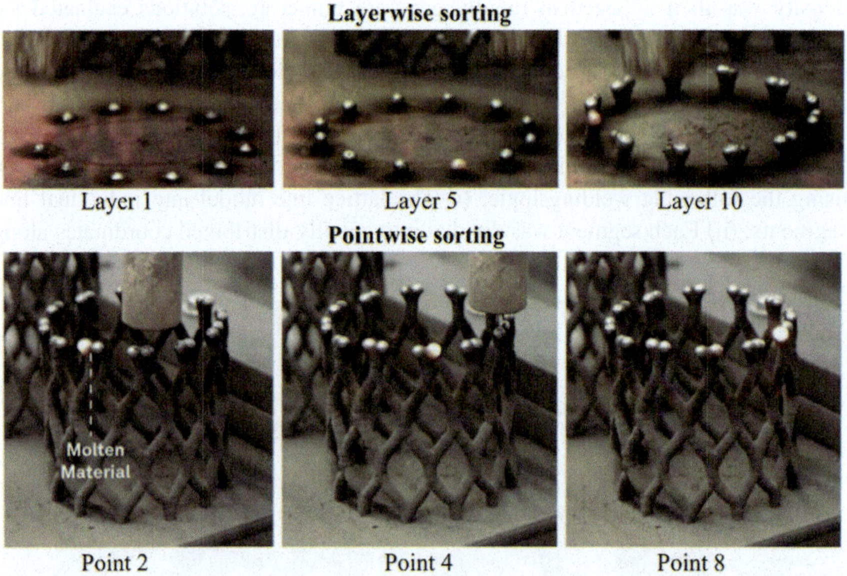

Fig. 7 The lattices were printed following two printing sequences. Firstly, points were ordered in a layerwise fashion according to their Z-Coordinate value. For each layer, points were ordered radially

The welding process used to produce the specimens generally resulted in good surface quality, however some errors occurred during fabrication. The most common issue was the welding electrode getting stuck in a solidified metal droplet, which caused the welding process to stop. This led to material discontinuities in the welded structure as well as an increased fabrication time since the welded surfaces and welding torch had to be cleaned, and the process restarted. As can be reasonably expected, material discontinuities also resulted in discrepancies in uniaxial compression testing results. Specimens which appeared to have no material imperfections behaved more favorably, since the structure seemed to behave as a coherent whole. This is seen by the global-buckling failure mode of the specimens (Fig. 8—specimen 02). On the other hand, specimens which either had inferior surface quality and material discontinuities had a reduction in performance of around 15%. Moreover, the specimens also tended to undergo a localized buckling failure around the area of material discontinuity, which suggests that the structure was not behaving as a continuous whole. This can be seen in specimens 01 and 03 in Fig. 8.

In conclusion, the study demonstrated the effectiveness of WAAM in producing lattice structures with stable cross-sectional dimensions. The parametric modeling approach allowed for rapid generation of numerous solutions which could be quickly prepared for fabrication. However, the study also highlighted the importance of the need for in-process monitoring and response to welding errors, as these had a considerable effect on the structural performance of the specimens.

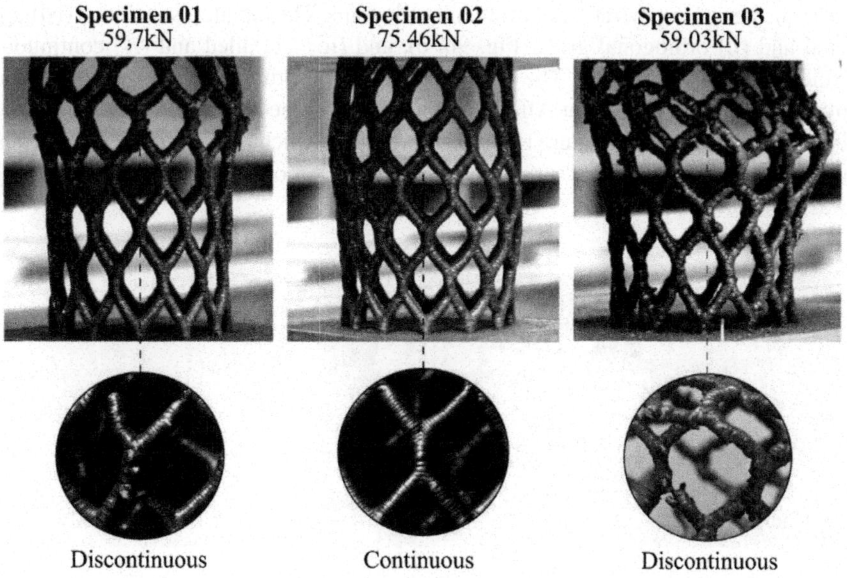

Fig. 8 The quality of welds had a noticeable effect on the performance of the structures. Specimens 01 and 03 exhibit roughly 15% reduction in strength compared to Specimen 02

3.3 *Function Integration: Reinforcing and Detailing Thin Sheet Metal*

Sheet metal is a widely used material in facade construction due to its properties such as malleability, strength, and durability. Advancements in digital design tools and sheet forming techniques have made it easier to create free-form facades using the material, which can be seen in notable structures like the Experience Music Project by Frank Gehry architects and the Morpheus Hotel by Zaha Hadid Architects. However, the drawback of using free-form sheet metal in facade construction is that it can make the fabrication and planning of the facade substructure more complex, despite sheet metal forming techniques like Multiple Point Stretch Forming being cost-effective. As a result, free-form facades are often rationalized into planar sheets of metal to achieve an overall free-form appearance.

The research project proposes a hybrid fabrication technique in which facade connection and reinforcing details are directly printed onto free-form sheet metal using WAAM. The aim of the research was to investigate the effects of various welding parameters and strategies on the distortion of sheet metal, and to then develop design strategies for reinforcing and adding connections to curved sheets of metal. The study explored design strategies such as intermittent welding (Fig. 9a), which involves producing short welds with gaps to reduce heat input and distortion, in addition to using the Fronius CMT cyclestp process that minimizes heat generated during welding. This led to the development of different welding patterns, which would later form the basis for generating reinforcement-layouts. These patterns included (i) diagonal and (ii) orthogonal grids (Fig. 9b, c), and (iii) Curtailed and (iv) continuous welding seams (Fig. 9d, e). The effects of different welding patterns were evaluated by (i) the extent of the Heat Affected Zone (HAZ)—the region surrounding a weld that experiences a temperature rise during welding, which is identified by the formation of color bands- as well as (ii) the overall deformation of the plates after welding was completed.

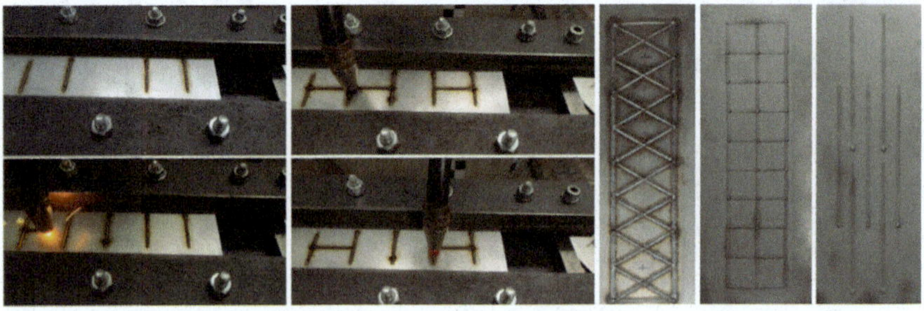

a b c d

Fig. 9 **a** Intermittent welding involves dividing welding lines into shorter segments. This led to the development of different welding patterns including **b** diagonal and **c** orthogonal grids as well as **d** curtailed seams

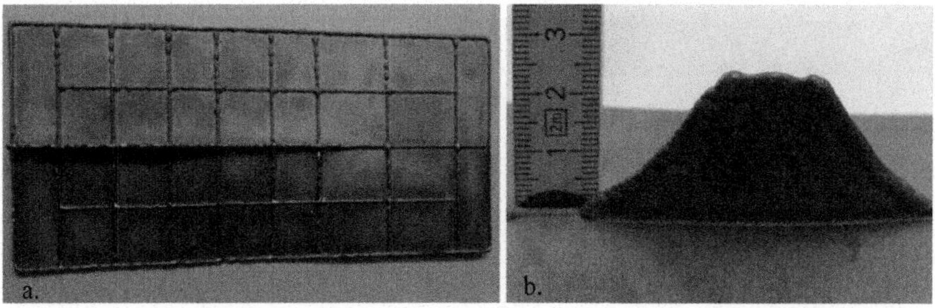

Fig. 10 **a** Welding an underlying grid pattern stiffened the sheet material which allowed for building up more layers of material, as seen by the central rib **b** It was also possible to build up multiple layers of welded material to form connection details

As could be reasonably expected, the effects of the spacing and lengths of welded seams had a visible effect on the plate material after welding. In particular, it could be seen that the increasing of the length of a weld seam had a negative effect on the overall distortion of the plate post-welding. Moreover, when it came to the buildup of multiple layers of weld material, the use of a diagonal and orthogonal grids aided in reducing the overall distortion (Fig. 10a). This could be due to the grid patterns increasing the overall bending stiffness of the plate and therefore, resistance to deformation. This also allowed for the buildup of larger connection details consisting of multiple layers of material (Fig. 10b).

Ultimately, this research aimed to add reinforcement and connections to freeform metal sheets for use in free-form façades. This was tested by pre-bending 0.75 mm gauge sheet metal into arbitrary forms. In order to maintain a stable welding process, a constant distance between the welding gun and the plates was needed. Therefore, a 3D-scanned model of the pre-bent sheet was needed. Many conventional scanning techniques require time-consuming post-processing of point-clouds. In order to overcome this, robotic tactile sensing was used to record a grid of pre-programmed points on the bent surface (Fig. 11). The resulting low-density point cloud allowed for the quick generation of a NURBS surface which was then used to recalculate welding paths based on the scanned model.

The use of WAAM to weld directly onto pre-bent sheet metal presents numerous exciting possibilities for constructing free-form facades. Figure 12a depicts a single stringer, roughly 1 cm tall, that was printed directly onto sheet metal. Figure 12b displays a single-layer welded curve network that represents the force-flow lines along the plate and was fabricated using an intermittent welding technique previously described. In the future, it is expected that the concepts presented in the two figures could be combined to create three-dimensional stringers or facade connections that are directly printed onto a curved plate, resulting in minimal material waste.

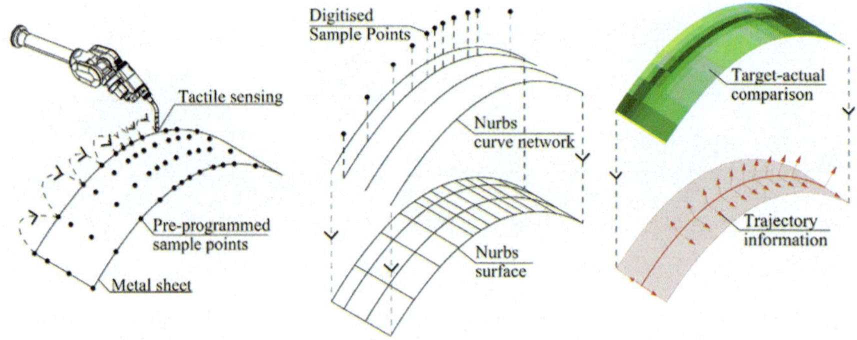

Fig. 11 Schematic workflow of using tactile (touch) sensing to record a predefined grid of points superimposed on an arbitrary surface. These points were used to create a NURBS curve network, which was converted into a NURBS surface. This information was used to generate new welding paths while also comparing the scanned and planned surfaces

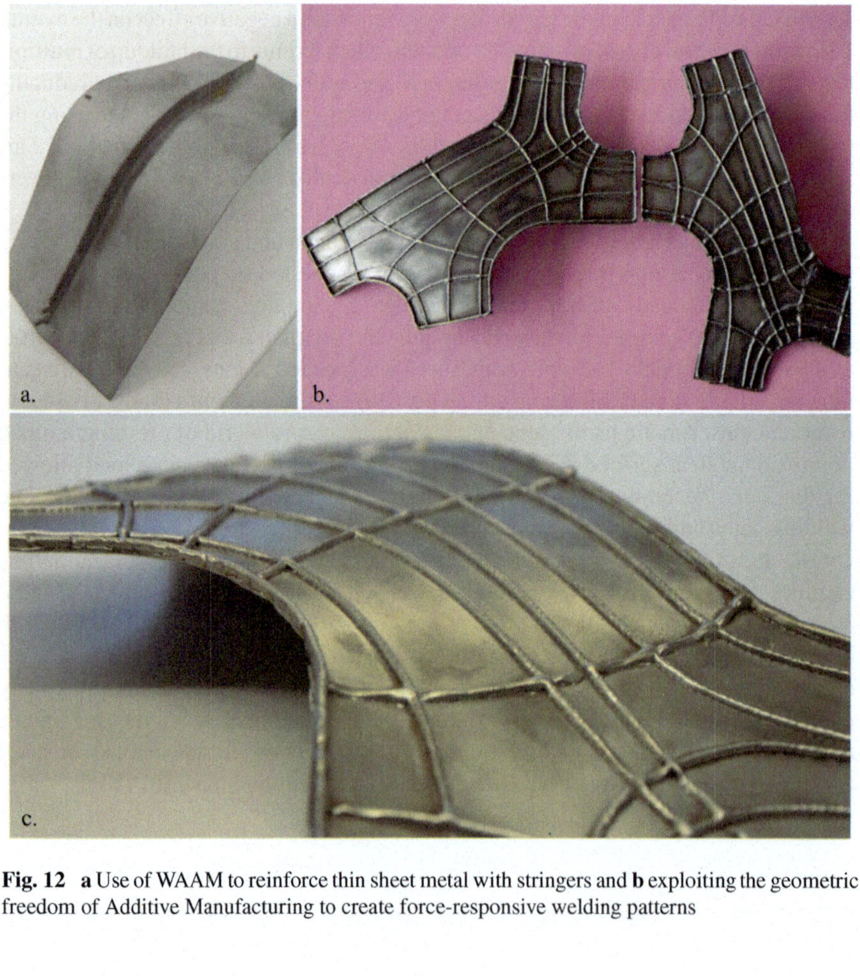

Fig. 12 **a** Use of WAAM to reinforce thin sheet metal with stringers and **b** exploiting the geometric freedom of Additive Manufacturing to create force-responsive welding patterns

3.4 Scaled Structure, Parametric Robotic Programming, Adaptive Process Control: Column Structures

In this paragraph, a research study is presented that focuses on combining digital design tools and Wire Arc Additive Manufacturing (WAAM) to produce exceptional columns that would be expensive to manufacture using conventional methods. In particular, the column structures show the use of a process referred to as Parametric Robot Programming (PRP), a method of defining robotic motion paths through mathematical functions alone. The selected columns are depicted as scale models in Fig. 13.

Model (a) is a column with a maximum edge length of 14 cm, whose rectangular cross-section uniformly increases or decreases in size as its height increases, reaching a total height of approximately 29 cm. Model (b) depicts an octagonal cross-section that experiences a constant twist over the element height, which suggests some possibilities for expressive architectural forms. Models (c) and (d) are columns with an ellipsoidal cross-section. In Model (c), both axes have been altered according to the moment loading of a doubly hinged column, with a corresponding imperfection profile. In this case, the structure has a smaller cross-section at the base, where lower bending moments are expected, and a larger cross-section near the middle. Model (d) was used to investigate the feasibility of (c) for more pronounced cross-section changes, particularly the manufacturability of the cantilevered form in the upper part.

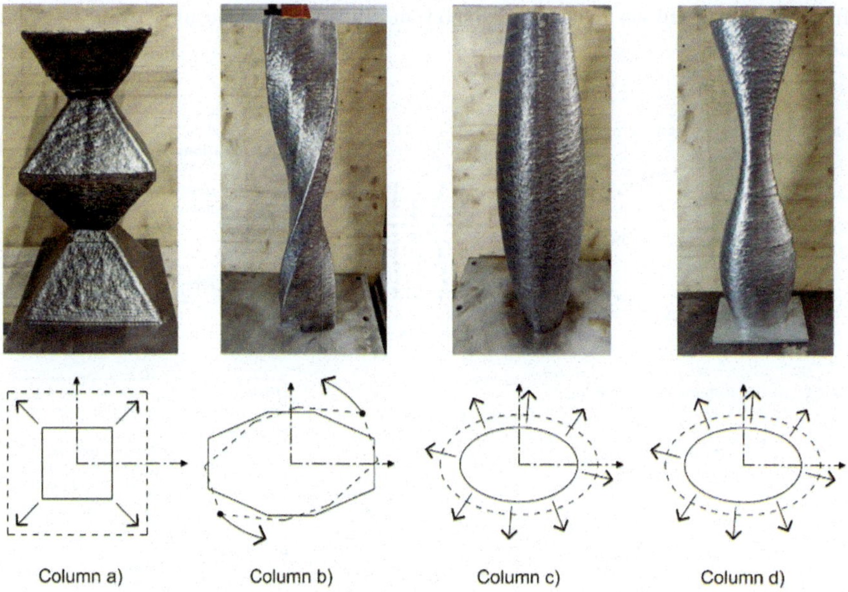

Fig. 13 **a** Investigated columns with a changing square cross-section, **b** a twisted octagon cross-section and **c, d** and two non-uniformly changing ellipse cross-sections (Lange et al. [24])

Columns (b)–(d) were fabricated to a height of approximately 50 cm with maximum cross-section widths of 11 cm for (b) and up to approximately 18 cm for models (c) and (d).

While the designs are relatively simple in terms of their geometry, they serve as first iterations towards more geometrically complex columns, which are defined by the limits of material deposition such as maximum overhangs, minimum cross-sectional areas, and wall thicknesses. This manufacturing approach could be combined with other design factors such as shape optimization to maximize load-bearing capacity.

Parametric Robot Programming: Defining the geometry through mathematical functions

Because of the relatively simple geometry, which could easily be described mathematically, a process referred to as parametric robot programming (PRP) was adopted. In conventional additive manufacturing, a three-dimensional object is sliced into individual layers, describing the motion path of a printer. In a PRP process, a structure is described entirely by mathematical functions. An example of how a code described using PRP is structured is presented as follows: Consider a column structure which is defined as an ellipsoid cross-section. Conventional slicing would require that (i) the entire structure is split up onto layers of predefined heights, and each layer is then (ii) divided into a few points through which the printer traverses to complete the printing of a layer (Fig. 14). This results in a high number of points being generated, with no room for adaptability should the height of layers have to change.

Now consider the same structure, which in turn is defined using mathematical functions. It is known that any point on an ellipse can be defined by the general equation:

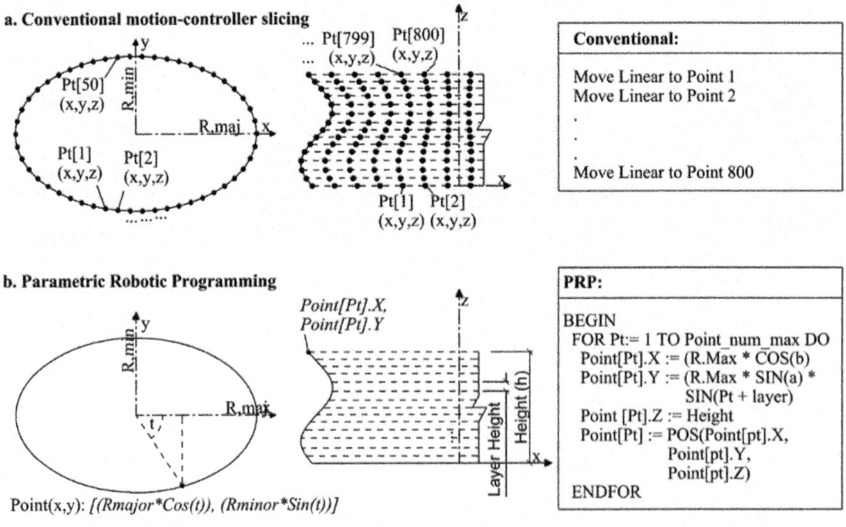

Fig. 14 **a** Conventional printing slicing **b** Slicing using PRP (Lange et al. [24])

$$\mathbf{Pt_{x,y}} = \mathbf{R_{max}} * \mathbf{COS(t)},\ \ \mathbf{R_{min}} * \mathbf{SIN(t)}.$$

where: $\mathbf{R_{max}}$ is the major radius of the ellipse

$\mathbf{R_{min}}$ is the minor radius of the ellipse

t is the angle at which the point is described.

By using this approach to define a shape, the general equations of the ellipse were uploaded to the robotic device, which calculated all the co-ordinates using the general equation given. By using functions to define the positions also allowed for more flexibility when there was a needed change in layer height. A generalized structure of the code which was used is summarized in the loop below:

```
BEGIN
   For Pt:= 1 TO Point_max DO
      Point (Pt).X := (R.Max * COS(t))
      Point (Pt).Y:= (R.Min * Sin(t))
      Point (Pt).Z := Zheight
      Point (Pt) := POS(Point(pt).X, Point(Pt).Y, Point (Pt).Y)
END FOR
```

As will be shown in the proceeding case studies, this approach of using mathematical functions as a means of slicing also allowed for dynamic change in layer height, as well as the generation of code for object with more geometric complexity.

3.5 Post Processing and Data Collection: Welding and Milling of Steel Nodes (WEMI)

In the Architecture, Engineering, and Construction (AEC) industry, steel nodes, which are the connections between structural elements in a building, are typically fabricated using traditional subtractive methods such as drilling and welding. This process involves cutting, grinding, and drilling steel plates, and then welding them together to form the node. However, this process can be inefficient and time-consuming. A significant amount of material is wasted during the manufacturing process, and manual welding can be labor-intensive and error-prone. The *WEMI* research project was an exercise in using WAAM to build up three-dimensional node structures. The structures were then post-processed using milling in order to account for dimensional inaccuracies arising from the WAAM process. Moreover, it addressed issues of in-process monitoring and control brought up in the previous case studies.

A 1:1 demonstration steel node (Fig. 15a) was designed to join different standardized steel profiles, namely circular- and rectangular-hollow-sections, as well as

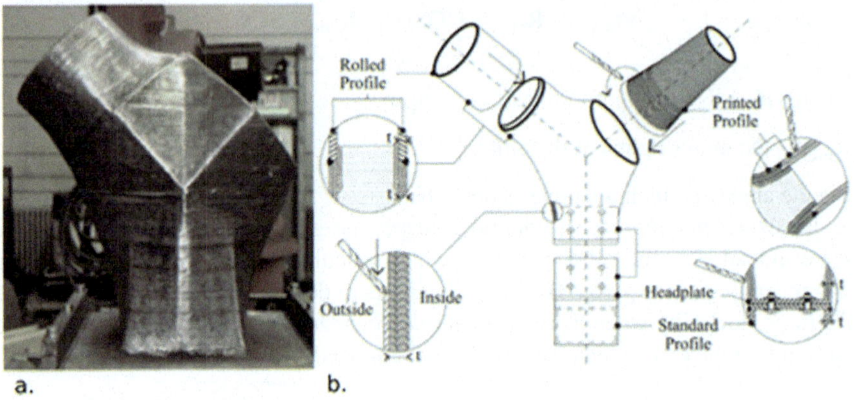

Fig. 15 **a** Welded node connection, prior to post-processing **b** Schematic of different applications and connection details for the steel node

printed ellipsoid-hollow-sections (Fig. 15b). This allows for cross-sectional profiles of different dimensions and geometric properties to be joined together using various connection techniques such as head-plates (figure bottom), plug connections with a welded overlap joint (left branch), or a welded butt joint (right branch). Furthermore, the process combines additive manufacturing with post-processing by removing material to achieve a planned wall thickness using milling and grinding processes.

Welding process feedback and control

As the print was around 50 cm tall, it was crucial to maintain a stable welding process to minimize the time spent on correcting errors and prevent a possible print failure. Adjusting robot motion paths, layer heights, and welding parameters is critical for a stable WAAM process due to potential discrepancies between the design geometry and actual printed geometry caused by differences in layer height or distortion during printing. Therefore, in-process data collection (and response) was essential to this project. A *rudimentary* digital twin was created by regularly storing values of the current welded layer height, as well as cooling, motion and welding times. By using a method of Parametric Robot Programming, a method of defining a layer by means of mathematical functions rather than predefined coordinates, real time feedback between the original 3D-model and actual printed geometry was possible. This allowed for a far more stable welding process, as robotic motion paths was regularly updated based on layer height information recorded with a frequent robotic tactile sensing to match the actual printed geometry (Fig. 16).

Post Processing

The same recorded data was later supplemented with 3D Scanning of the printed structure which was then used for generating the robot trajectories used for the milling post-processing step. In this case, a LIDAR scanner was used to generate a mesh geometry of the finished structure, and conventional slicing was used to obtain

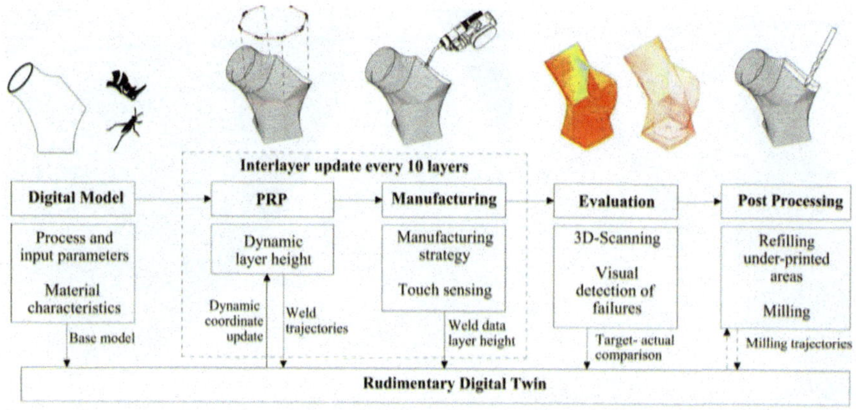

Fig. 16 Overview of generating a digital train using PRP (*Source* Waldschmitt et al. [24])

the trajectories. Milling was carried out on the premises of Spannverbund GmbH, using a KUKA KR-300 robotic arm fitted with a Tecknomotor milling device. The grasshopper 3D plugin, Kuka-prc, was used to communicate the milling trajectories to the controller. Milling was carried out using an overlap of 2 mm and an upwards-milling process. After the milling process was completed, flap sanders with grades ranging from 60–240 were used to achieve a smooth surface finish (Figs. 14 and 17).

3.6 Large Scale and Automated: Parametric Robotic Programming of a Steel Printed Bridge

One of the more promising applications of WAAM is its implementation in onsite fabrication. The *AM Bridge* was a research project carried out by TU Darmstadt in 2018. Figure 18 shows a schematic of the bridge fabrication, which was designed as two welded structures which gradually cantilevered over a small pond, eventually intersecting in the center. Although this approach allowed for printing the bridge directly on site, it brought with it a number of problems. Firstly, at the time of the project, welding parameters for cantilevered structures was an area of research which was still relatively unexplored. Moreover, the scale of the structure presented problems in the way in which data was generated and stored. Conventional slicing would have resulted in large amounts of coordinates which would have to be communicated to a robotic device which could only store up to 900 variables in a single file. This would have led to an excessive amount of individual files being needed, all of which would have to be manually uploaded to the controller. Moreover, conventional slicing did not allow for much room for flexibility should any errors occur during the manufacturing process. Therefore, the method of Parametric Robot Programming was also adopted for fabricating the structure [25].

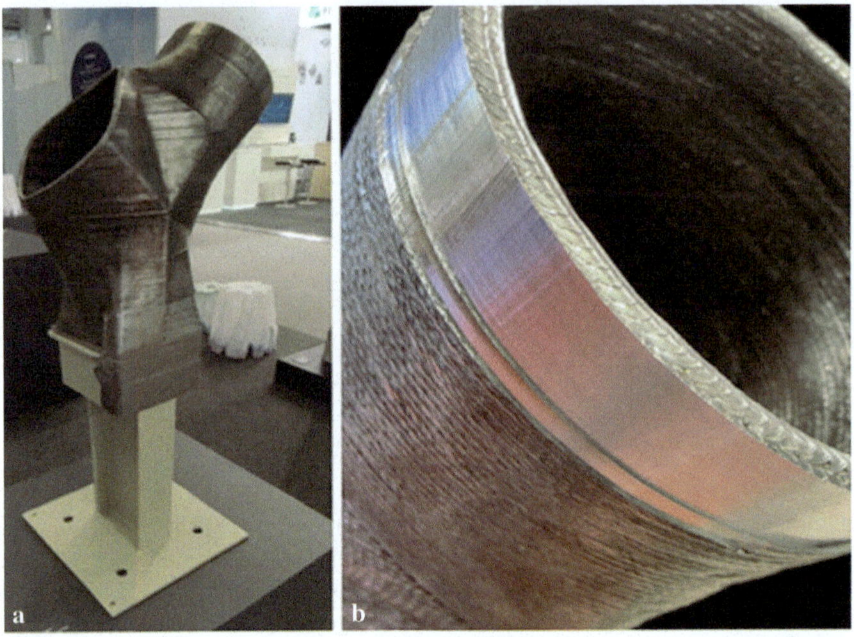

Fig. 17 **a** The final Steel node structure with ellipsoid connection in the foreground, mounted to rectangular base plate connection **b** Closeup of the post-processed circular hollow connection (*Source* Waldschmitt et al. [24])

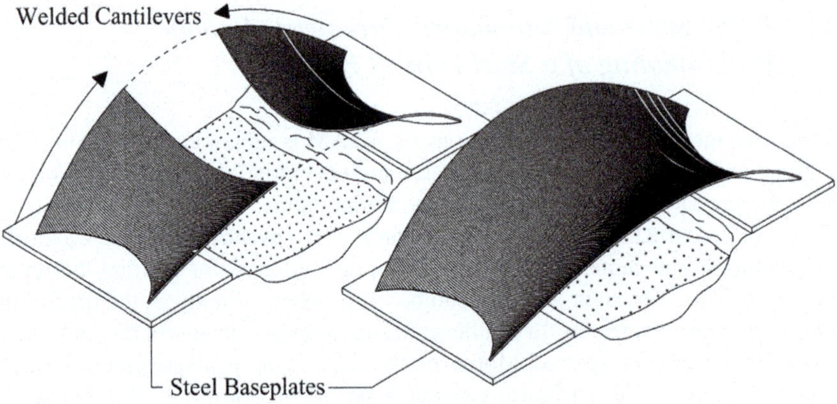

Fig. 18 Schematic of the *AM Bridge manufacturing*

Polynomial regression and Parametric Robot Programming (PRP)

Parametric robot programming uses mathematical functions to describe the structure's geometry, where x- and y-coordinates are functions of the z-coordinate, allowing for flexibility in reacting to deviations between target and actual geometry without explicitly giving the robot coordinates, while drastically reducing the amount of code needed ([24, 25]). The *AM Bridge* was designed using a NURBS modeling technique, and was therefore not defined by readily-obvious mathematical functions. To overcome this, the final bridge geometry was rationalized as a series of polynomial functions, which may be generally expressed as:

$$y_i = a_1 + a_2x_i + a_3x_i^2 + \cdots + a_n x^{n-1} \text{for } (i = 1, \ \ldots, \ n)$$

The polynomial functions were calculated using a polynomial regression function within Grasshopper3D and MATLAB. This is calculated by determining the best-fitting curve through a series of data points. In order to obtain the data points to perform polynomial regression, the bridge geometry was divided into a number of isoparametric-curves (Iso-curve) which were projected onto X–Z and Y–Z planes in order to obtain a two dimensional curve (Fig. 19).

For each Iso-Curve, the corresponding X–Z and Y–Z projections were divided into a number of points, which were spaced according to the curvature of the projection curve (Fig. 20a). Using Grasshopper 3D, the points were used as data input for performing polynomial regression, which obtained a polynomial function based on a user-defined expression degree. The polynomial functions were then used to reconstruct the original curves and compare for the deviations between the original (Fig. 20b). In order to reduce the complexity of the mathematical expressions (which

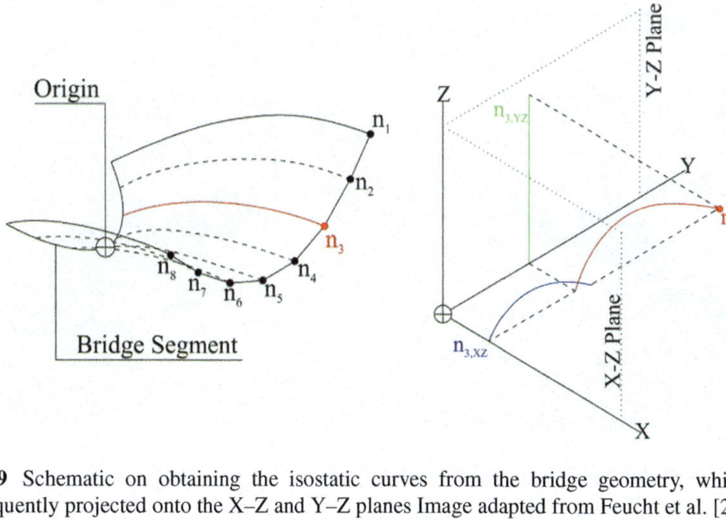

Fig. 19 Schematic on obtaining the isostatic curves from the bridge geometry, which were subsequently projected onto the X–Z and Y–Z planes Image adapted from Feucht et al. [25]

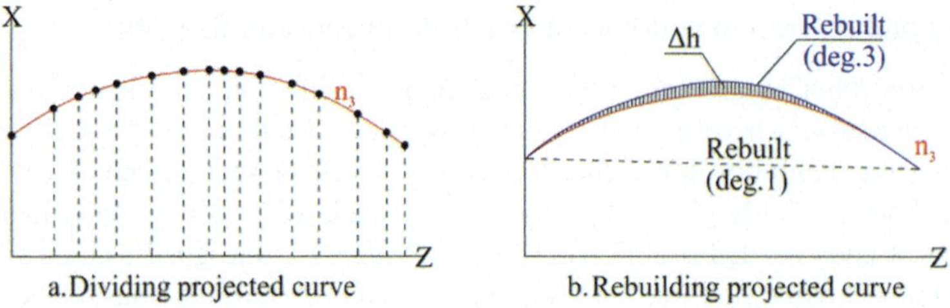

Fig. 20 **a** The curve projections are divided into a number of points, which serve as data points for polynomial regression. **b** Polynomial regression is used to obtain polynomial functions and curves which are compared against the original. Image adapted from Feucht et al. [25]

would have to be solved by the robotic device), the results with the lowest degree and acceptable deviation from the original curve were chosen.

This approach allowed for a very flexible method of printing; during the welding process, Robotic touch sensing was performed regularly to measure the Z-Coordinate values of the layers. Because the layers were defined by mathematical expressions in terms of the Z-coordinate value, this allowed the robot trajectories to be altered according to what was measured. Such a flexible approach would not be possible using standard slicing in which there is no possibility of updating coordinates based on recorded data.

Unfortunately, despite the efforts in material research for cantilevered structures as well as the establishment of an efficient slicing method based on PRP, the bridge structure was not completed. This was due to the effects of thermal distortion of the large, and slender, cantilevers during manufacturing which caused the two segments to twist out of plane. An initial fix to this was the inclusion of a shorter center strip which connected the two cantilevers together, concluding the end of the manufacturing (Figs. 21 and 22).

Fig. 21 The final bridge structure which was printed onsite, over a shallow pond

Fig. 22 The final bridge exhibited at FORMNEXT 2019

4 Comparison and Discussion of Case Studies

The case studies presented in the previous served as a snapshot of some of the recent advancements in WAAM within the AEC. While they reveal AEC applications which are quite applicable in the near future, they also shed light on certain aspects of WAAM which should not be overlooked.

Process control and adaptability

One of the themes which was re-occurring in many of the case studies was that of in-process monitoring and adaptation to what was happening. This was highlighted in the *Welding and Milling (WEMI)* project, enabled using Parametric Robot Programming (PRP). As evidenced by other projects, such as the *3D Printed Lattices* which exhibited a reduction in structural performance due to welding failures, the use of PRP and in-process monitoring holds great benefits for ensuring a high-quality production process. Developments Image-based detection tools could also be beneficial in tackling these issues. In 2019, researchers at Virginia Tech developed an image-based closed loop system for the live monitoring, defect detection and process parameter adaptation of additive manufacturing with Fused Filament Fabrication (FFF). In the case of WAAM, this approach of live-altering welding parameters combined with the adapting of layer heights already exhibited in PRP, could lead to far greater control of the welding process. An interesting case is also put forward by the researchers at the Royal Danish Academy of Fine Arts, who acknowledge the need for prediction modeling of additive material as it changes state during production and put forward proposals for Machine Learning-Controlled fabrication processes.

The theme of process control extends to what was presented by the *AM Bridge* project—onsite fabrication. The concept of onsite additive manufacturing isn't a new one; WIlliam Urschell showed us that it was possible to produce concrete structures like this as far back as the 1940s [26]. With WAAM, however, the margin of error during manufacturing is not as forgiving as that of concrete, with minor deviations potentially adding to fabrication time or, in the worst case scenario, the conclusion of a project. This exact issue came to light during the *AM Bridge* project, where the unpredicted thermal distortion of the manufactured structure ended up cutting the project short.

Performance enhancement and function integration

A second theme we see is that of using WAAM as a means of enhancing the performance or adding additional functionality to standardized components. In the *Reinforced Sheet Metal* project, both functionality and performance is introduced into deformed sheet metal material. Functionality is introduced by printing facade connection details directly onto the sheet metal, and performance is enhanced through the addition of welded stiffeners. We see this theme of using WAAM to enhance standardized connections also in other applications. Feucht et al. [18] for example, used WAAM to directly stiffen IPE sections, as did the researchers of TU Braunschweig ([19]). This same concept is also exhibited by Stelia Aerospace who used WAAM to stiffen aluminum plates. This brings forward the question of whether or not WAAM (or AM) should be seen as a means of completely replacing existing structural components, or should we be striving for obtaining the best of both worlds? The *WEMI* project, for example, does this by using WAAM only where it is needed, that is, at the complex intersection between standardized profile sections.

Mass customization

In the context of freeform Facade construction, which typically involves numerous components ranging from hundreds to thousands, it's natural to consider the concept of mass customization when discussing WAAM. Considering the importance of design flexibility, analysis, and information flow within this context, it's clear that the process of mass customization can significantly impact the design process. This was seen in both the *Spiderglass* and *Reinforced Sheet Metal.* In the former project, this effect was on the simplification of the printed geometry in favor of one which was quick to digitally produce and fabricate. In the *Reinforced Sheet Metal* project, this impact also resulted in the adoption of a curved-based approach to designing and analyzing the structural components.

Acknowledgements The authors would like to thank Thilo Feucht M.Sc and Maren Erven M.Sc, for their help in the setting up and running the WAAM Lab at TU Darmstadt. Specific project acknowledgements are:

Spider Glass Connections: Prof. Dr-Ing Ulrich Knaack, Dr. Ing.- Christopher Borg Costanzi (ISMD, TU Darmstadt), Dr.-Ing Jörg Hildebrand and Matthias Eiber M.Sc (TU Ilmenau), Prof. Dr.-Ing Andreas Eichhorn and Dr.-Ing Florian Schill (Geodesy, TU Darmstadt), Dipl. Ing. Martin Manegold and Dipl. Ing. Dominik Zausinger (Imagine Computation GmbH) and Dipl. Ing. Jochen Hölscher (Hölscher GmbH).

Lattice Structures: Ihab Said M.Sc, Dr. Ing.-Christopher Borg Costanzi M.Sc.

Welding and Milling (WEMI): Prof. Dr.-Ing Jörg Lange, Benedict Waldschmitt (IFSW, TU Darmstadt), Prof. Dr.-Ing Ulrich Knaack, Dr.-Ing Christopher Borg Costanzi,Dipl.-Ing Thomas Engel, Dipl.-Ing Jan Müller, Prof. Dr-Ing Stefan Böhling (Spannverbund GmbH) Dr.-Ing Andreas Eichhorn and Dr.-Ing Florian Schill (Geodesy, TU Darmstadt).

AM Bridge: Thilo Feucht M.Sc, Maren Erven M.Sc, Benedikt Waldschmitt M.Sc, Dr.-Ing Christopher Borg Costanzi M.Sc, Prof. Dr.-Ing Jörg Lange, Prof. Dr.-Ing Ulrich Knaack.

The Spiderglass and Welding and Milling projects were part of the Building AM Network and supported by Jöckel Innovation Consulting GmbH. The research project is funded by the "Bundesministerium für Wirtschaft (BMWi)" and "Zentrales Innovationsprogramm Mittelstand (ZIM)".

References

1. Snooks (2022) Behavioral tectonics: agentBody prototypes and the compression of tectonics. Arch Intell 1(1):1–14. https://doi.org/10.1007/s44223-022-00007-8
2. Butterfly Screen (2021) MX3D. https://mx3d.com/industries/art/butterfly-screen/. Last Accessed 27 Feb 2023
3. Lunar floor for ESA (2021) MX3D. https://mx3d.com/industries/construction/lunar-floor-for-esa/. Last Accessed 27 Feb 2023
4. MX3D Bridge (2021) MX3D. https://mx3d.com/industries/infrastructure/mx3d-bridge/. Last Accessed 27 Feb 2023
5. Wire Arc Additive Manufacturing (WAAM), https://www.twi-global.com/technical-knowledge/job-knowledge/arc-based-additive-manufacturing-137.aspx. Last Accessed 28 Feb 2023
6. Jin W, Zhang C, Jin S, Tian Y, Wellmann D, Liu W (2020) Wire arc additive manufacturing of stainless steels: a review. Appl Sci 10(5):1563. https://doi.org/10.3390/app10051563
7. Wire arc additive manufacturing of aluminium alloys: a review (2021) Mater Today Proc 41:1139–1145. https://doi.org/10.1016/j.matpr.2020.09.153
8. Lin Z, Song K, Yu X (2021) A review on wire and arc additive manufacturing of titanium alloy. J Manuf Process 70:24–45. https://doi.org/10.1016/j.jmapro.2021.08.018
9. Steel-copper functionally graded material produced by twin-wire and arc additive manufacturing (T-WAAM) (2022) Mater Des 213:110270. https://doi.org/10.1016/j.matdes.2021.110270
10. Full-Service Metal Additive Manufacturing for Manufacturers, https://www.waam3d.com/industries. Last Accessed 28 Feb 2023
11. STELIA Aerospace develops 3D printed demonstrator for fuselage panels (2018) https://www.aerospace-technology.com/news/stelia-aerospace-develops-3d-printed-demonstrator-fuselage-panels/. Last Accessed 28 Feb 2023
12. Ariza I et al (2018) In place detailing. Combining 3D printing and robotic assembly. In: Proceedings of the 38th annual conference of the association for computer aided design in architecture (ACADIA), Mexico City, Mexico. https://doi.org/10.52842/conf.acadia.2018.312
13. Laghi V, Palermo M, Gasparini G, Trombetti T (2020) Computational design and manufacturing of a half-scaled 3D-printed stainless steel diagrid column. Addit Manuf 36:101505. https://doi.org/10.1016/j.addma.2020.101505
14. Erven M, Lange J (2022) Design of optimized 3D-printed steel nodes. In: Structures and architecture a viable urban perspective? CRC Press, London, pp 229–236. https://doi.org/10.1201/9781003023555-28
15. LASIMM, https://lasimm.eu/gallery-video.html. Last Accessed 28 Feb 2023
16. Integradde (2021) MX3D. https://mx3d.com/services/research/integradde/. Last Accessed 28 Feb 2023
17. Erven M, Lange J, Feucht T (2021) 3D-Printing with steel of a bolted connection. Ce/Pap 4(2–4):825–832. https://doi.org/10.1002/cepa.1367
18. Lange J, Feucht T (2019) 3-D-printing with steel: additive manufacturing of connection elements and beam reinforcements. In: IABSE symposium, Guimarães 2019: towards a resilient built environment risk and asset management, Guimarães, Portugal. https://doi.org/10.2749/guimaraes.2019.1836

19. Kloft H, Schmitz LP, Müller C, Laghi V, Babovic N, Baghdadi A (2023) Experimental application of robotic wire-and-arc additive manufacturing technique for strengthening the I-Beam profiles. Buildings 13(2):366. https://doi.org/10.3390/buildings13020366
20. BE-AM Map (2021) BE-AM. https://be-am.de/be-am-map/. Last Accessed 28 Feb 2023
21. Krypton, a column in Aix-en-Provence-XtreeE, https://xtreee.com/en/project/krypton/. Last Accessed 28 Feb 2023
22. 3D Printing Climbs to the Top at Chattanooga Airport Chattanooga, TN (2022) BRANCH TECHNOLOGY. https://branchtechnology.com/3d-printing-climbs-to-the-top-at-chattanooga-airport-chattanooga-tn/. Last Accessed 28 Feb 2023
23. Purple S.T.E.A.M (2021) BRANCH TECHNOLOGY. https://branchtechnology.com/purple-s-t-e-a-m/. Last Accessed 28 Feb 2023
24. Waldschmitt B, Lange J, Costanzi CB, Knaack U, Engel T, Müller J (2022) Robot supported wire arc additive manufacturing and milling of steel columns. In: Current perspectives and new directions in mechanics, modelling and design of structural systems, 1st edn. CRC Press, pp 361–367. https://doi.org/10.1201/9781003348443-59
25. Feucht T, Lange J, Erven M, Costanzi CB, Knaack U, Waldschmitt B (2020) Additive manufacturing by means of parametric robot programming. Constr Robot 4(1):31–48. https://doi.org/10.1007/s41693-020-00033-w
26. Concrete Without Forms (2017), https://www.youtube.com/watch?v=nkxe9uK8jSo. Last Accessed 28 Feb 2023

Dr.-Ing Christopher Borg Costanzi completed his doctoral thesis at Darmstadt Technical University in the field of Wire Arc Additive Manufacturing for freeform façade construction. He completed his master's degree at Delft Technical University, graduating with Honors and Summa Cum laude, where he researched non-conformal concrete additive manufacturing. He currently works at the design and construction team at BMW Munich (Via. Brunel GmbH).

Benedikt Waldschmitt is a Doctoral candidate at the Institute of Steel Construction, Darmstadt Technical University. He is currently researching the use of Wire Arc Additive Manufacturing as a fabrication technique in the steel construction Industry. He also completed his master's degree at the Institute of steel construction, where he researched cantilevered welded structures and steel bridge fabrication in Wire Arc Additive Manufacturing.

Chapter 13
Custom-Made Printheads Empowering 3d Printed Concrete: Innovations in Digital Design and Fabrication of Complex Prefabricated Elements

Volker Ruitinga and Emma Chris Avramiea

Abstract Over the past decade, robotic manufacturing has made its way out of the shadow of the automotive industry, giving rise to a growing interest in adopting Additive Manufacturing (AM) techniques for architectural and infrastructure projects. This chapter presents the explored capabilities of 3D concrete printing (3DCP) as developed by Vertico, an innovative startup from the Netherlands that focuses on additive manufacturing with concrete. It explores the role that the symbiosis between robotic technology, material mix and computational design tools is playing within the larger context of the construction industry, in prompting material manufacturers to create mixes better suited for 3d printing, and technology manufacturers to focus on advancing the tools to mix and extrude seamless layer after layer of mortar. What differentiates Vertico's technology from other concrete printers is their two-component (2 k) technology and perseverance in pushing the limits of concrete printing. Vertico does not focus on printing entire buildings, and certainly not on-site. It creates building blocks for a revolution taking place at the intersection between architecture and robotics. This revolution, for the designer as well as for the user, means an unprecedented freeing from building conventions, which are generally guided by strategies such as cost cutting, production limitations and increasing standardization.

Keywords 3D concrete printing (3DCP) · Robotic manufacturing techniques · Computational design · Topological optimization · Additive manufacturing · Industrial automation

---

V. Ruitinga · E. C. Avramiea (✉)
Vertico, Eindhoven, The Netherlands
e-mail: emma.chris.avramiea@gmail.com

V. Ruitinga
e-mail: info@vertico3d.com

P. Ruttico (ed.), *Coding Architecture*, Digital Innovations in Architecture, Engineering and Construction, https://doi.org/10.1007/978-3-031-47913-7_13

1 Introduction

1.1 First Steps in 3d Concrete Printing

Vertico was founded in 2018, a propitious year for concrete 3D printing. The municipality of Eindhoven, in collaboration with the local Technical University (TUe) had just announced the plan to deliver five concrete 3D printed houses [1].

The ingredients of Vertico's entry in the 3D printing industry were: a fascination for additive manufacturing (AM) techniques, a great passion for experimentation, and robotics experience borrowed from the automotive industry. While its first 3D printing techniques were still being developed, Vertico embarked on a project of high complexity, whose goal was to 3D print a small house. This house, whose roof's design proved to be the biggest challenge, was realized in collaboration with the consortium Huizenprinters, Jelle Feringa and Summum Engineering. Continuous iteration of machines and software resulted in the development of a process for strongly accelerating the hardening of concrete. Such technological developments in turn allowed the printing of highly complex designs. Setting a high bar in terms of the designed objects' geometry and structural requirements had a positive snowball effect on Vertico's technology developments and fabrication strategies (Fig. 1).

In the following years, computational and concrete AM techniques continued to evolve worldwide. Advancements in material science are crucial to concrete printing.

Fig. 1 Dutch Design Week columns developed in collaboration with TU Eindhoven

With the Eggshell project, developed during a Master Thesis at the TU Delft University [2], set-on demand concrete enables the utilization of a very thin 3d printed formwork. This technique is a precedent of the lost formwork technique that the 3D printing industry advanced. Projects such as the Striatus bridge by Block Research Group (BRG) realized in collaboration with Zaha Hadid Architects (ZHACODE) [3], advanced the status quo of concrete manufacturing, by combining 3D printing with masonry techniques. The Striatus bridge is made of structural elements that are dry-assembled without any form of additional reinforcement, while reducing concrete use by up to 70%, in an industry with a significant carbon footprint. In this context, Vertico embraces the concrete 3D printing scene to set bold new standards in terms of material structure and aesthetics.

1.2 Computational Design, Robotic Fabrication Industry

The current chapter presents the evolution of 3DCP at Vertico from different perspectives: (a) the computational tools as instruments for design and toolpath generation; (b) the manufacturing advantages; (c) the robotic technology and its adoption in concrete additive manufacturing with the help of mortar pumping systems.

The innovation described here is that of a two component (2 k) 3DCP technology allowing for on-demand setting of concrete. Finally, this chapter will address Vertico's role within the larger construction industry and the challenges faced at the printing scale that the startup operates (Fig. 2).

As of 2023, Vertico is a global leader in providing cement 3D printing solutions, with clients across Europe, the United States, Eastern Asia, The Middle East and Australasia. In addition to installing state-of-the-art robotic printing systems, Vertico prints projects, designs software solutions for robot and gantry printers, and provides consulting on matters of 3D concrete printing.

(a) (b)

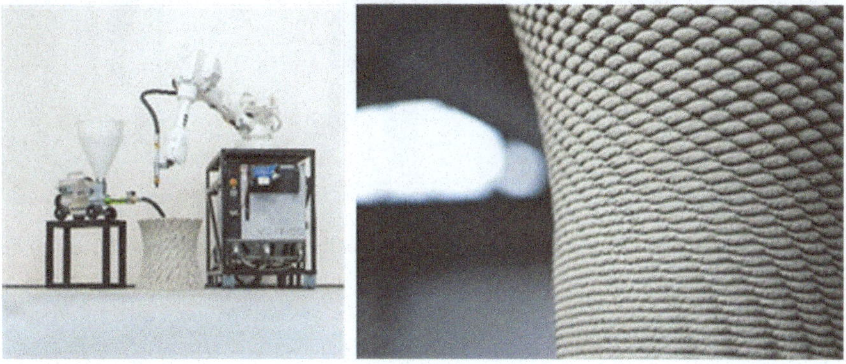

Fig. 2 **a**, **b** Vertico small-scale printing solution. Print pattern

1.3 Long Term Vision

The global shift of focus towards economies of scale has resulted in the shaping of the modern architectural movement. The architectural styles corresponding to the industrial movement resulted in a departure from human touch of the building trade. Robotic manufacturing offers an opportunity to revive the level of craftsmanship achieved by architects such as Gaudí with the Sagrada Familia, by introducing technologies and design tools that could reduce costs and make them competitive with current building technologies. Moreover, similar or even higher levels of detailing achieved through additive manufacturing processes could highly optimize the production time and precision [4], as well as reduce the need for laborious and costly craftsmanship.

New business cases have been uncovered with the development of AM for concrete, associated with the unlocking of a new aesthetic. This is where Vertico's initial vision arose from, a vision to augment the cultural value of building construction with the aid of automated manufacturing techniques. For such architectural artifacts, their high cultural and material value will help them stand the test of time and in doing so maximize their value (Fig. 3).

Fig. 3 Column for the Dutch Design Week, realized in collaboration with TU Eindhoven

2 Content

2.1 Augmented Potential of Computational Design

Parametric Design in Architecture

In its essence, parametric design is an umbrella of tools which render design decisions quickly adjustable through the use of input parameters. As common in other robotic manufacturing circles, Vertico develops its computational design software through the plugin Grasshopper for Rhino. While designing the toolpath that will be utilized for printing a column for instance, the coder can modify its height, diameter, convexity, rotation angle of a pattern along the column's length, thus making it customizable on a multitude of levels that could be incorporated in the script. While the software does not discriminate between styles or design approaches, the use of parametric techniques does often produce a recognizable aesthetic, further complimented by the formal freedom permitted by additive manufacturing (Fig. 4).

One of the early projects designed by Vertico with the aid of computational tools in Grasshopper is the Voronoi fence wall. This project was realized in collaboration with Saxion Industrial Design Research Group, De Witte van der Heiden Architecten and Trebbe, and manufactured from individual pieces, each with a maximum weight of fifty kilograms. This dimensional constraint allows for stacking the wall's fifty pieces by only two operators. When the dimensional requirements changed throughout the

Fig. 4 Fence design realized based on Voronoi algorithm in Grasshopper

design phase, the use of the Voronoi algorithm enabled adjusting the design without having to go through lengthy re-drawing, mainly by adjusting the input parameters.

Besides the new aesthetic potential, AM is a suitable technique for building *strength through geometry,* since it enables material to be printed only where structurally necessary. Grasshopper can be utilized to perform structural analysis on geometry, which in turn informs design decisions. The Voronoi fence wall showcases how computational design can create intricate objects even when only using basic (1 component) printing technology.

Topological optimization

Topological optimization is a popular concept in industries where a high strength-to-weight ratio is critical, such as aerospace, medical, automotive and construction. The optimization model stems from a mathematical algorithmic method that iteratively analyzes and removes redundant material parts. In computational design circles, it is defined as optimizing the material use in relation to the structural function to be performed. Being a close parallel to self-organization processes of matter in the natural world, the results of topological optimization materialize as organic shapes. Their curved configurations are cost prohibitive if produced with regular manufacturing techniques or formwork. Additive manufacturing is uniquely suited to produce objects with a high degree of variation and curvature.

One of the applications of topological optimization that Vertico developed is for the design of a bridge. For ease of on-site assembly, infrastructure elements are often designed and constructed out of prefabricated elements. With the aid of concrete 3d printing, Vertico realized a topologically optimized bridge made out of unique elements, which were assembled after curing. The project was realized in a collaboration led by the University of Ghent and Technion Israel Institute of Technology (Fig. 5).

Besides the original computational basis behind its morphology, this bridge also constitutes its own formwork. Upon assembly, the printed parts are cast and post tensioned. In this context, the fragment becomes a secondary manufacturing method, one priorly integrated, as formwork to the ‘cast bridge’.

Freedom over toolpath design—customized software

In order to give physical shape to a digital 3D model, the path of communication from human designer to robot is through code. The designer’s intention of obtaining an intricately curved object can only be deciphered by the robot’s computer when formulated as an ordered list of coordinate positions. Besides parametric design conceptualization, one of the most useful applications Vertico has for using Grasshopper is that of toolpath design and translation into code readable by the robot’s computer. The innovative start-up has created its own toolpath generating software, which can import a 3D model created by the client, translate it into an optimal toolpath for printing and return an XYZ coordinate code for the robot. For Vertico the toolpath creation is an opportunity to experiment with intricate designs and develop printing strategies resulting in high quality detailing and finishing. The direct translation from 3D model to coordinate code represented lowering the technical (robotic) knowledge

Fig. 5 Topologically optimized 3DCP bridge

threshold necessary in 3DCP for computational designers, empowering them with the tools to operate robotic arms. Nonetheless, a practical understanding of material behavior for concrete 3D printing is crucial for the design process. Once the designer has built a sound intuition of the cycle where design and manufacturing inform each other, this knowledge can be standardized and re-utilized.

2.2 Robotic Printing Techniques

As exemplified by the topologically optimized bridge, one of the key values of robotic manufacturing is the morphological design freedom it enables. In particular as regards 3DCP, additive manufacturing brought about a new formal language and structural solution-space for concrete. Some of the most demanding challenges in concrete 3d printing that Vertico has explored with are: non-planar printing (HPA Pavilion, concrete canoe), overhangs of up to 60° (column experiments), absence of formwork, and achieving heights of 2.5+ meters in one print with the support of its accelerated technology.

Another aspect of the progress made throughout time with Vertico's technology includes the development of a material mix that sets quickly. This is crucial for 3D printing with cement, since the stability of the print depends on the stability of each layer. While in traditional concrete construction, the cement can have higher water content and longer setting time, due to being supported by a formwork until it has set,

Fig. 6 **a**, **b** Canoe printed out of cement with up to 90° overhang

the mortar mix utilized in concrete printing needs to be engineered for its particular application. To facilitate quick setting time for the extruded material, at Vertico, the robot printing system enables the mixing of two components, the cement mix and an accelerant (Fig. 6).

Non-planar printing

Besides enabling quick stacking of layers, the fast setting property of the mortar also enables layer deposition at angles different from horizontal. Since 3D printed objects are created through layer by layer deposition, the mechanical properties of the prints depend on the direction of printing. The resulting prints are not isomorphic, like cast concrete prototypes, and perform much better in compression than in tension [5]. Printing in a direction perpendicular to the applied forces, even when the printing direction is variable, enhances the structural properties of the print by allowing the transfer of the compressive loads in the optimal direction [3].

One way to achieve variable layer height is by incrementally changing the printing speed and taking advantage of simple fluid dynamics—the faster the robot arm moves, the less material is extruded over the distance traveled. Since less material is extruded, less space is necessary to deposit it, so the distance between extruder and print can be lower, thus resulting in a thinner layer overall. Alternatively, when the speed is lowered and the nozzle brought higher, more material is extruded, thus resulting in greater layer height. All variables resulting in differential layer height are part of a calculated prediction and integrated by the designer in the toolpath geometry and printhead behavior (speed, orientation) (Fig. 7).

What this relatively simple maneuver enables is the printing of non-planar layers. Starting from the bottom layer and building upon differential thickness throughout (Fig. 8).

The HPA pavilion is an example of Vertico's early experimentation with non-planar printing. Inspired by lightweight concrete flooring systems [6] and designed as a modular structure, its fabrication did not involve the use of formwork. Also, for the assembly, no casting was performed, nor any custom formwork made. Instead, the parts were bound together with a thin layer of mortar, thus applying masonry

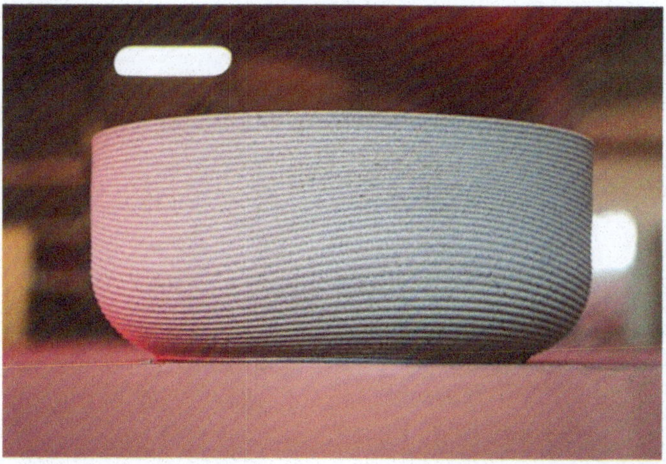

Fig. 7 Variable layer height print

techniques to 3d printed elements [3], which resulted in a highly materially-efficient structure.

Lost formwork

In cast concrete projects, the geometry of a concrete element is achieved by pouring concrete in a form that has the shape of the desired final result. The formwork is subsequently removed upon setting [2]. 3DCP enables the elimination of the need for a removable structure. When casting is utilized in such scenarios, the 3D printed part is integrated within the structure and is not subsequently removed.

The economical advantages of the lost formwork technique are that it eliminates material waste and reduces post-processing labor. One of Vertico's early projects, the topologically optimized bridge, is an example of experimentation with lost formwork, where casting and rebar were integrated in the printed formwork to increase the structural strength.

Combining the morphological freedom of 3D printed elements with the lost formwork technique could be the key to advancing the use of additive manufacturing for architectural and infrastructure projects of large scale.

Rebar + infill

Although through their design, 3D printed concrete elements can be optimized for structural bearing purposes, the novelty of the technology and its unique structural properties represent a challenge in the way of standardization of 3D printed elements. The construction industry relies on internationally defined standards for construction, certainly within the European context.

(a)

(b)

Fig. 8 **a, b** HPA Pavilion. No permanent formwork

Compared to a standard concrete column, a 3D printed column does not display isotropic properties. The layer-by-layer deposition of material, with the layers typically distributed horizontally, makes 3D printed concrete elements perform better under compressive loads and be more fragile under tensile stress.

For increasing its performance to tensile loads, 3DCP elements can be equipped with rebar and infilled with concrete (Fig. 9). However, introducing rebar makes it more difficult for concrete to be recycled; structurally, infill increases the weight and thus generates additional stress. So far, Vertico has attempted to focus on the

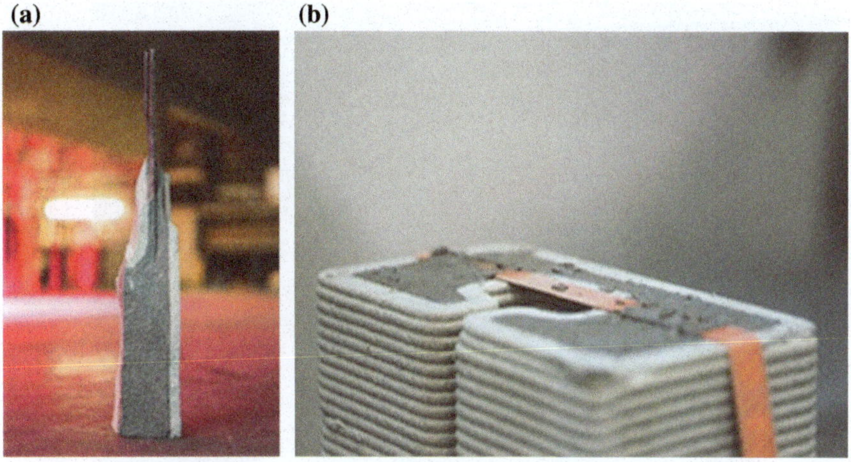

Fig. 9 a, b Column reinforced with rebar. b Concrete infilled column

more sustainable option, creating strength through geometrical design, minimizing the necessity for additional reinforcement.

Strength in compression

See Fig. 10.

2.3 The Robot System

Adapting technologies from the automotive and construction industry, Vertico has come up with a unique printing system solution, currently utilized worldwide in research facilities as well as by commercial parties from various industries. This printing system is the core of the start-up's continuous improvement efforts.

The Vertico concrete printing system combines an industrial robot arm with six-degrees of freedom on a six meter track, a state-of-the-art mortar pump, accelerant dosing pump and proprietary Accelerator Printhead (Fig. 11).

Two Component (2 k) system—accelerated technology

A result of multiple technological iterations, the robotic system currently utilized at Vertico provides an accelerated 3DCP solution. In terms of hardware, this sub-system is composed of the robotic arm with the extruder printhead, as well as an accelerant pumping station. The two components (2 k), the cement and the accelerant solution are mixed into a homogeneous solution at the level of the printhead, right before being extruded.

This accelerated cement mix allows a shorter time gap between the printing of a layer and the next one, with a setting time of less than thirty seconds. The cement cures

Fig. 10 Urban furniture project loaded with 1,000 kg of cement as strength test

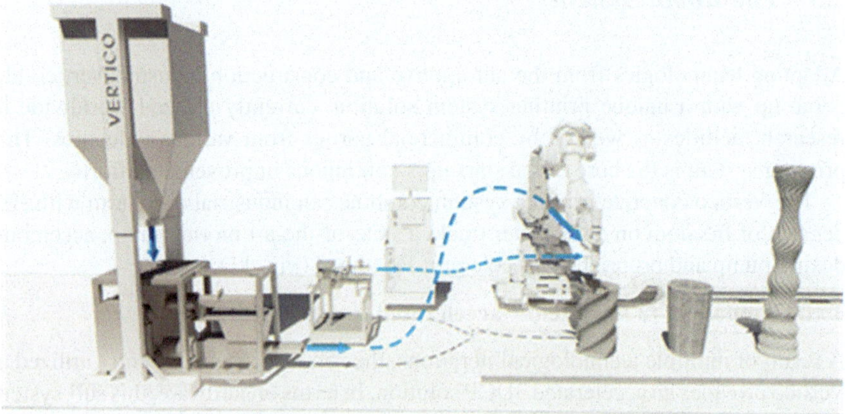

Fig. 11 Vertico 3DCP system with: Mortar pump (MAI), silo, accelerant station (Vertico), robotic arm (ABB), Accelerator Printhead (Vertico), Control and Live Monitoring (Vertico)

enough in the meantime to support its own weight and the one of subsequent layers without considerable deformation. This stability allows for printing taller structures with much greater levels of intricacy (Fig. 12).

Most 3DCP applications for architecture focus on printing straight walls, which is the more conventional application of concrete printing. The added value of accelerated concrete in terms of printing capabilities relies on the higher geometrical and structural complexity achievable, in the form of overhangs, designing strength

Fig. 12 Nozzle system that accelerates the cement mix

Fig. 13 Example of overhang stability due to set-on demand concrete technology

through geometry or morphological freedom. As the Vertico system builds on conventional concrete printing, it is more than capable of doing both standard and accelerated concrete printing (Fig. 13).

Automation

The robot utilized by Vertico incorporates a number of sensors for the monitoring of activities around the production and extrusion of cement. The data collected through a multitude of sensors was integrated in an in-house designed monitoring software, for visualizing any fluctuations and to provide useful feedback regarding the printing parameters. Being able to visualize and store data is highly important in understanding the relationship between extruded material quality and the qualities of the printed product, as well as obtaining feedback regarding the operational accuracy of the machines involved, in particular the cement pump. Moreover, this monitoring system represents a step closer to the automation of the 3D printing process with cement.

Owing to their impact on the cement quality and print curing, two main variables are of interest here: temperature and pressure. The pressure of the mortar is measured as it exits the pump and as it enters the printhead. The temperature of the mix is measured both before and after the mixture is accelerated, because of the thermal reaction resulting when accelerating the mix. In addition, temperature is also measured at the pump. Finally, the temperature and humidity of the environment in

Fig. 14 Installation of robot track

which printing occurs are measured and recorded, since these values have an effect on the rate of water loss through evaporation from the printed object.

Track

The standard XL robot printing solution that Vertico currently operates with consists of a robot placed on a track, which moves on one axis over the distance of five meters. This allows printing of objects with a length of up to seven meters (Fig. 14).

Material

Due to the nature of the 3D printing process, the performance requirements and qualities of 3D printed cement are different from the ones of cast concrete. The extrusion process results in a cementitious mix which is considerably more porous than traditional concrete. To make it pumpable, the water content for such a mix needs to be relatively high. Because it is laid without formwork, each layer also needs to be structurally self-sustaining, thus requiring a mix that sets fast enough for the addition and support of subsequent layers.

While the ultimate goal is to collaborate on the research and provision of the material mix with a knowledgeable global industry partner, so far Vertico has predominantly utilized its own material cement mix. With a list of clients interested in the robotic system distributed worldwide, having a reliable, locally sourceable material is the sensible alternative for setting the base for a sustainable industry. The collaboration with global material manufacturing industries has proved to be informative for both sides, with 3D printing industry actors such as Vertico performing tests and setting the pace towards the creation of high-performance materials especially designed for 3D concrete printing.

2.4 Industry

In its role as expert on additive manufacturing with concrete, Vertico has collaborated through the years with various entities, including but not limited to academia, architectural studios, urban artists, private clients and actors from various other industries. It continues to develop, sell, install and upgrade concrete printing systems worldwide. The current subsection will expand on an ongoing project developed in collaboration with the European Space Agency and the Technical University of Delft.

Collaborative research

The project follows the participation by TU Delft researchers in a call for proposals launched by the European Space Agency (ESA) in 2020, the '*Off Earth manufacturing and construction*' Campaign. The project consists in the proposal for a human habitat design on Mars, whose construction would be performed with a fully autonomous swarm of rovers, laying material additively on site [7].

Vertico's contribution as industry partner began with testing the design proposals by manufacturing small scale prototypes. The geometry of the habitat is based on a three-dimensional Voronoi logic, where for manufacturing purposes, every wall has an inclination higher than 45°, to prevent failures due to overhangs.

Additionally, Vertico has incorporated regolith simulant in the material mix. Regolith, a fine planetary rock and dust mix, is a resource widely available on the Martian surface. Due to the high costs of Space Exploration missions, regolith is a superiorly more sustainable in-situ resource than transporting materials from Earth for construction purposes. Experiments performed with regolith simulant have proven that it is a feasible alternative to replace aggregate in cementitious mixes, but more research is necessary to formulate a highly reliable material mix (Fig. 15).

The exploration of construction scenarios in extreme environment conditions once again highlights the advantages of robotic additive manufacturing, especially the qualities of being programmable, automatable, fast and risk-mitigating.

2.5 Challenges

Even with rapid development over the past two decades, 3DCP is a relatively novel manufacturing technique that brings about a number of challenges to tackle. During setting, through the loss of water content in the mix, concrete can display shrinkage behavior [8]. If the shrinkage occurs at a fast rate, the internal forces between the particles of concrete can be superseded, resulting in different types of cracks on the concrete surface. The rate of the loss of humidity in the cement depends highly on

Fig. 15 3d printed iterations of the Mars habitat with regolith simulant

the environment in which curing happens. Thus, controlling the quality of the prints directly translates also in controlling the environmental conditions for printing in the initial stages.

Buckling failure is another challenge of 3DCP that occurs mostly during printing. It can be detected as a gradual deformation. This deformation usually becomes greater as the weight increases, such as when new layers are printed on top. The main cause for buckling is the lowered capability of bottom layers to support the weight of subsequent layers, causing a domino effect. To tackle this issue, the consistency of the mix and the concentration of accelerant solution in the mix need to be adjusted.

Print failures can also be tackled by building material robustness. Currently the success of a print is dependent on fine-tuning parameters of the print to a narrow range of tested values. Increasing material robustness would result in a material that performs well even when there is a change in parameters such as batch, temperature, water concentration or dosing variation of accelerant. However, designing material robustness is a challenging task in itself.

The alternative to building material robustness is perfecting standards for 3DCP printing. This translates into having a deep understanding of the precise combinations of parameters that guarantees printing quality and ensures repeatability.

2.6 Concerns on Sustainability

When designing structures with high cement content, strategies to reduce the material mass utilized, such as topological optimization and elimination of formwork, are key features for sustainability. Topological optimization, as previously addressed, is a mathematical method for lowering the material content utilized for the task, while maintaining or improving weight-bearing capacity. Using no formwork is a complementary sustainable action since it eliminates the need for material that would support the structure until it has set.

Ultimately, increasing the durability of a project's lifetime is a key aspect of sustainability. The longer the life span, the higher the value of the energy spent in creating it. Designing and printing customized elements with high aesthetic and structural value are highly relevant for validating their irreplaceable value.

In the future, Vertico intends to work on sustainable material alternatives for cement. 3D concrete printing can be separated into four main categories: robotics, software, pumping and material. Currently, Vertico excels in robotics and software, with extensive experience and a strong partner in pumping in MAI International GmbH. Having excelled in the other areas of 3DCP will enable Vertico to expand its work on sustainable material alternatives for which it stands already open to. Currently Vertico is involved in developing a new material mix with Sika, but the robotic system works with material from most current suppliers.

3 Conclusion

3.1 Future Outlook: From Initial Vision to Developments in 3DCP

The aim of design has always been and remains to make the world a more beautiful place. The longevity of those things that we bring into the world and the care with which we maintain them are closely correlated. Many of the structures that continue to inspire us today have long since surpassed their intended lifetime. The trend however has been one of cost reduction, standardization and minimalism. With 3D printing there is a unique opportunity to reintroduce a more divergent aesthetic at competitively low costs. In other words, we can make beautiful things whilst keeping costs under control.

Monje

Monje is a spinoff brand launched by Vertico for manufacturing 3D printed planters. It grew as a continuation of early toolpath experimentation with columns. Monje brings design customization to everyday life objects, presenting a new aesthetic of concrete manufacturing. As every other project at Vertico, also Monje is a testbed for experimenting with the capabilities of 3D robot printing, in particular with the

Fig. 16 Planters from the Monje collection with varied printing patterns

aesthetic of overhangs and various printing patterns, as well as layer stability (Figs. 16 and 17).

Technological innovation

The giant upon whose shoulders Vertico stands is its innovative robotic technology. It enables the weaving together of unique works of digital art materialized in cement-based mortars. The craft is considerably less labor intensive than analogue manufacturing. In the spirit of Kaizen, implementing small incremental improvements to one of the components of the printing process has proven to be the key to increasing printing quality, both aesthetically and structurally.

The need for automation in the construction industry remains high. The added value of a worker has actually decreased over the last half a century. Yet the demand for construction is and will continue to remain high. In order to realize this demand, innovative construction technologies are essential [9] (Fig. 18).

Directions in Design

In the five-year span that Vertico has established itself and printed with regularity, the advancements in terms of robotic technology and material mix have enabled successful prototyping of challenging projects. Prints with overhangs up to 60°, non-planar printing, achieving structural strength through geometry are only a few of the technical challenges resolved. In the process, the exploration with toolpath design aesthetics has opened possibilities of an architecture that inhabits a more

Fig. 17 Planters from the Monje collection with variable color injection

organic language, controllable through parameters, and simultaneously improving the sustainability of concrete utilization (Fig. 19).

Fig. 18 Concrete Canoe with colored pigment injection

Fig. 19 3D printed concrete chaise-longue

References

1. Lojanica V, Colic-Damjanovic VM, Jankovic N (2018) Housing of the future: housing design of the fourth industrial revolution. In: 2018 5th International symposium on environment-friendly energies and applications (EFEA). IEEE, p 3
2. Burger J, Lloret E, Taha N, Scotto F, Demoulin T, Mata-Falcón J, Gramazio F, Kohler M, Flatt R (2020) Design and fabrication of a non-standard, structural concrete column using eggshell: ultra-thin, 3D printed formwork. https://doi.org/10.1007/978-3-030-49916-7_105
3. Bhooshan S, Bhooshan V, Dell'Endice A, Chu J, Singer P, Megens J, Van Mele T, Block P (2022) The Striatus bridge: computational design and robotic fabrication of an unreinforced, 3D-concrete-printed, masonry arch bridge. Arch Struct Constr. https://doi.org/10.1007/s44150-022-00051-y
4. Dillenburger B, Hansmeyer M (2013) The resolution of architecture in the digital age. International conference on computer-aided architectural design futures. Springer, Berlin, pp 347–357
5. van den Heever M, du Plessis A, Kruger J, van Zijl G (2022) Evaluating the effects of porosity on the mechanical properties of extrusion-based 3D printed concrete. Cem Concr Res 153:106695. https://doi.org/10.1016/j.cemconres.2021.106695
6. Hawkins W, Orr J, Tim I, Shepherd P (2017) Developing an innovative lightweight concrete flooring system for sustainable buildings. In: IABSE symposium report, p 108. https://doi.org/10.2749/222137817821233233
7. Robotic Building homepage, http://www.roboticbuilding.eu/project/rhizome-development-of-an-autarkic-design-to-robotic-production-and-operation-system-for-building-off-earth-hab itats/. Last Accessed 20 Dec 2022
8. Lerch W (1957) Plastic shrinkage. Proc ACI J 53(8):797–802
9. The Economist homepage, https://www.economist.com/business/2017/08/19/efficiency-eludes-the-construction-industry. Last Accessed 20 Dec 2022

Volker Ruitinga is passionate about 3D printing, digital manufacturing and architecture. He founded Vertico in 2018. His previous work experience was in the automotive industry in finance, project- and value chain management. He holds a master's degree in Philosophy (cum laude) from the Erasmus University.

Emma Chris Avramiea has developed her passion for Computational Design at Politecnico di Milano, where she obtained her Bachelor of Science in Architecture, followed by a Master in Computational Design and Robotic Manufacturing at the TU Delft. During her Master, she participated in exciting collaborations with both the space industry as well as the robotic manufacturing industry. In her role as Quality Assurance Engineer at Vertico, she helped set the first steps towards the standardization of concrete 3D printing for the company.

Chapter 14
Architects as Innovators: Harnessing the Power of Digital Design and Construction

Lucas ter Hall

Abstract The shift towards digital design and construction in architecture seems to be omnipresent and irreversible. The opportunities that come along with this digital evolution appear to be vast but still sub optimally used by different stakeholders within the realization of architecture. This opens the opportunity to connect the data these specialists create in one single, digital seamless flow of data describing the complete lifecycle of a real estate project. It seems that architects could play a pivotal role in this shift. There may be a central role and as such the opportunity to add extra value in the creation of a streamlined digital workflow while touching a more fundamental issue, namely how architects can strengthen their position from which they can claim and protect their creativity and aesthetic concerns under capitalist conditions. In this article these thoughts are excavated and discussed by introducing some of the work of Studio RAP.

Keywords Digital fabrication · Computational design · 3d printing · Ceramics · Timber construction · Robotics

1 Opportunity for the Architect

1.1 Introduction

Rapidly evolving digital technology brought endless new possibilities to the way we envision, design, and construct the built environment in the last few decades. Many specialists within the AEC-industry adapted digital tools to manage, organise and execute construction tasks. They digitised previously analog tasks which substantially improved workforce productivity and extended their technical capabilities to make the impossible possible. This opens the opportunity to connect the data these specialists create in one single, digital seamless flow of data describing the complete

L. ter Hall (✉)
Studio RAP, Scheepsbouwweg 8–K10, 3089 JW Rotterdam, The Netherlands
e-mail: lucas@studiorap.nl

P. Ruttico (ed.), *Coding Architecture*, Digital Innovations in Architecture, Engineering and Construction, https://doi.org/10.1007/978-3-031-47913-7_14

lifecycle of a real estate project. The promise of such digital workflow is to set project specific objectives as early as the initialising stage and keep track of them during the design, engineering, construction, maintenance, dismounting and remounting stage. Potentially enabling a truly holistic approach to a circular life cycle of a building with a new set of contemporary aesthetics.

1.2 The Creation of Data

Since the rise of digital computers, stakeholders within the AEC-industry started to digitise the data they created. This started the transition from a purely analog building process to a semi-digital supply chain. Semi-digital, as most contemporary building processes can be framed as digitally scattered, where many specialists combine analog tasks with digital tasks for different clients with different needs, responsibilities and interests. Without a transcending idea about how digital data should be exchanged between stakeholders of a building process, the unique benefits of a complete digital workflow are not taken to an advantage. Most of the time there's no stakeholder within a contemporary building process that takes full responsibility on this matter for the complete life cycle of a building. The key in realizing this is to follow a streamlined protocol on how to exchange purely digital data to communicate and execute project specific objectives from the very beginning of a project till the very end of project. In such a digital lifecycle everything that's digital needs to stay digital, from early project specific objectives till the reuse of every building part.

The digital road from the initiative phase of a building project till the demounting and reconfiguration of building components still needs to be constructed and will be fairly complicated. To realize this still many different specialists are needed to fulfil all building regulatory and client specific requirements. Tender laws in Europe make this scattered building process even more profound. This matter asks for a stakeholder that sets up a protocol for the digital workflow of a project, but for which stakeholder could this be an opportunity?

1.3 The Digital Caretaker

Most of the time there's no digital caretaker available during the building process as most clients are not familiar with the broad opportunities a complete digital process can deliver. These days most consultants are adopting digital tools, but there's one consultant that has a historically close relationship with the initiator and the end user of a project which is connected to a project from an early development stage, the architect. As most architects are digital-native these days, it's obvious that streamlining this digital workflow could be an opportunity for this specialist. Also, in the light of the eroded role of most architects which have lost much ground in recent years this can be fairly interesting. In the publication Why Architects Matter [1], Flora Samuel explains this diminishing role. Where project objectives related to costs and benefits took over the more subjective approach to the design of the built environment which includes matters like social, cultural, contextual and aesthetic considerations. This thinning role of architects doesn't stop at the profession itself but has long lasting effects on people's life, as they undergo the outcome of the architect's work every day.

The benefits of becoming such a digital caretaker can be twofold for the architect, they can play a crucial role in creating a streamlined digital workflow and reclaim a more central role in the building process. This last benefit touches even a more fundamental issue, because it gives architects the opportunity to strengthen their position from which they can claim and protect their creativity and aesthetic concerns under capitalist conditions.

In this perspective Studio RAP is explicitly aiming at reclaiming such a significant role for the architect in the designing and realization process of the built environment. To show how such a central role in the designing and realization in a building process could work, three case studies are discussed where the role of the architect went beyond their traditional responsibilities. These three case studies are the Main Concert Hall of Theatre Zuidplein (Rotterdam, NL), Poortmeesters|New Delft Blue (Delft, NL) and Ceramic House (Amsterdam, NL).

2 Main Concert Hall of Theatre Zuidplein

The main concert hall is part of the interior of Theatre Zuidplein built on the South bank of Rotterdam (NL) and was delivered in May 2020. The acoustic walls which in one fluid movement embraces the audience and the artist were designed and realized using a complete digital workflow. The project was designed with the use of algorithms, which resulted in a rippling ocean of thousands of triangles, creating the perfect acoustics possible for this venue. Studio RAP functioned in this project as a digital caretaker and initiated a team of (digital) specialists, set up the protocol for exchanging digital data, digitally designed the acoustic walls, generated digital production data and ensured a digital database with the realized building parts for future maintenance and reuse (Fig. 1).

2.1 Introduction

The City of Rotterdam commissioned Studio RAP to design and realize the interior of the main theatre hall within a tight budget and tight schedule. Because the original tender phase didn't give a solution that could be realized within time and budget the client was looking for a different approach. For that reason, the municipality came across the Rotterdam-based architecture firm that is pioneering the integration of digital techniques in the construction process, aiming to realize projects with a

Fig. 1 Picture of the Main Concert Hall of Theatre Zuidplein (P. Top photographer)

complete digital workflow. With this innovative approach, by digitizing the wall design from start to finish with self-developed algorithms, the project was realized within budget and planning (Fig. 2).

Fig. 2 Close up picture of the Main Concert Hall of Theatre Zuidplein (P. Top photographer)

2.2 *Digital Workflow*

All the phases of this project have been realized fully digitally and parametrically, from design to realization. As the theatre was almost built the project could start from a digital point cloud. Through the algorithmic design of Studio RAP, engineering firm Arup made its acoustic and structural calculations and Studio RAP generated only digital production data for the contractor Aldowa. The advantage of this method is that everything remains digital and therefore variable till the production and execution phase. This opened up the possibility to make changes without any problems until late in the realization process. In this way, different design scenarios can easily be tested for planning, technical and financial feasibility. Studio RAP functioned as the digital caretaker in this project ensuring the right digital data was created, used and exchanged.

2.3 *Optimised Sound*

Together with Arup over 10.000 virtual theatre hall variants were simulated. The hall was, as it were, digitally kneaded after which the best variant was chosen. Using advanced software, calculations were made to measure the effect of a certain curvature in the design on the reflection of the sound of a speaker or musician on the stage. Afterwards this main shape is translated into smaller triangles which in turn are adjusted. Where so-called ‘pure reflections’ are desired, the area of triangles is flat. Where diffuse reflections are desired, the area of triangles form a folded surface. The result is an even distribution of sound across the entire auditorium, so that every member of the audience can fully enjoy the show or concert they came to see. The walls don’t just affect the auditorium’s soundscape: they also break the light in countless color tones inside the theater to create a warm and dynamic play of colors. Because each triangle is folded at a unique angle, the light is refracted in different shades of red (Figs. 3 and 4).

3 Poortmeesters|New Delfts Blue

3.1 *Introduction*

Another project where Studio RAP played a central role within the realization is Poortmeesters|New Delft Blue, a pioneering project in the historic Dutch city Delft. Also for this project Studio RAP executed tasks that go beyond the traditional architects’ consultancy, as they were also responsible for the contracting and 3d-printing of this unique project. This project reinterprets the world-famous decorative qualities and design vocabulary of Delft Blue porcelain. By fusing 3D clay printing,

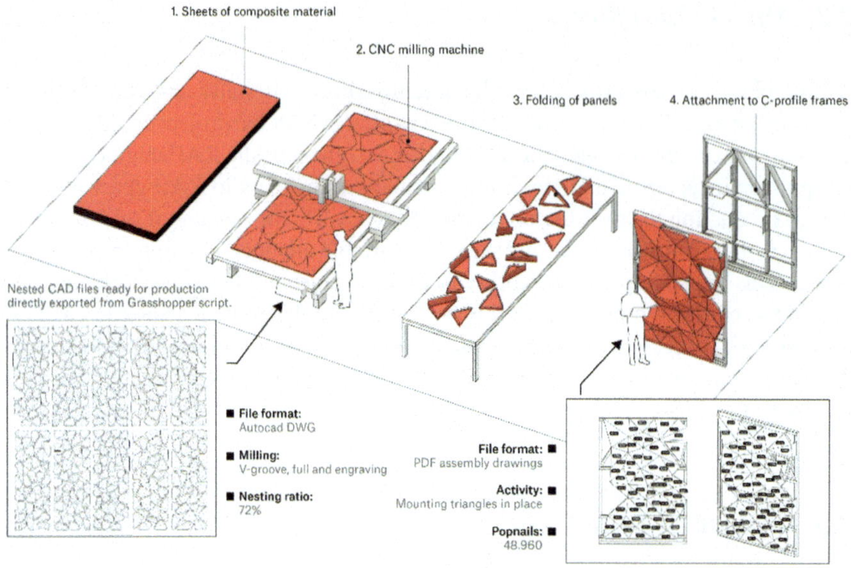

Fig. 3 Illustration of the fabrication process the Main Concert Hall of Theatre Zuidplein

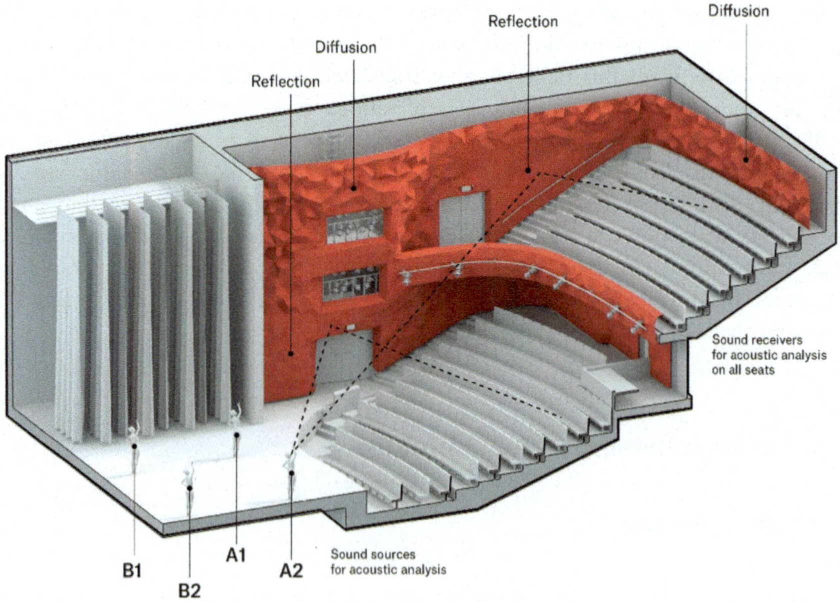

Fig. 4 Illustration of the sound reflection of the Main Concert Hall of Theatre Zuidplein

computational design, and artisanal glazing, New Delft Blue hopes to unfold a new architectural potential of ceramic ornamentation of the twenty-first century.

3.2 Design Vocabulary

The lush communal courtyard of the Poortmeesters building block, which is part of the large-scale redevelopment Nieuw Delft, is framed by two large entry gates. These refer to Delft as a medieval trading city with its many city gates and provide passers-by with a glimpse of the idyllic life going on inside. The New Delft Blue gates take their inspiration from traditional Delft Blue porcelain plates. The design vocabulary of these plates has changed radically over the centuries from copying Chinese and Japanese Oriental designs to making contemporary Dutch designs. The differentiation between a decorative, abstracted frame and picturesque scene, however, has been a recurring trait. As such, this division is the starting point for the New Delft Blue design, where the scene is the ever-changing courtyard life while the gates stand as a three-dimensional frame that focuses the view (Fig. 5).

Fig. 5 The New Delft Blue project inside Poortmeesters building (R. De Vecchi photographer)

3.3 New Delft Blue

The gates, four meters wide, eight meters high and twelve meters deep, cover a large public staircase. The ceramic cladding is designed to balance the two functions of the gates: indicating a separation between public space and communal space, while also guiding one's attention to the courtyard. The deep blue color reflects the connection to Delft Blue porcelain and the surrounding canals, and it provides a pleasing contrast to the earth-toned brickwork of the building. A serpentine line in the ornamental pattern guides viewers' eyes past the straight lines of the staircase and the gate toward the welcoming scene at its center.

The approximately 3.000 tiles that covers the two gates are 3D printed, allowing for the creation of contemporary ceramics that are both rich and unique. Using an algorithmic approach to 3D pattern design, certain manufacturing constraints—maximum overhang, width, height and depth, shrinkage constraints, and internal support structure—were taken into account when the geometry of the tiles was generated. Because the tiles are 3D printed, applying variations to their shapes allow for poetic ways of "painting with shapes." This is done, for example, by applying a transparent blue runny glaze on the convex portions of the tiles (the hills) while allowing for deep blue glaze pools in the concave (valley) areas of the tiles. This allows for smooth transitions between hues of blue that are impossible to produce otherwise. 3D-printed-ceramic is a high-quality cladding material that offers low maintenance and high durability in even the most corrosive environments, i.e. deserts and coastal areas. As such, the potential for 3D-printed ceramics goes far beyond the applications within Delft and for this reason it will also be applied in a design by Studio RAP for two residential high-rise projects at this moment (Fig. 6).

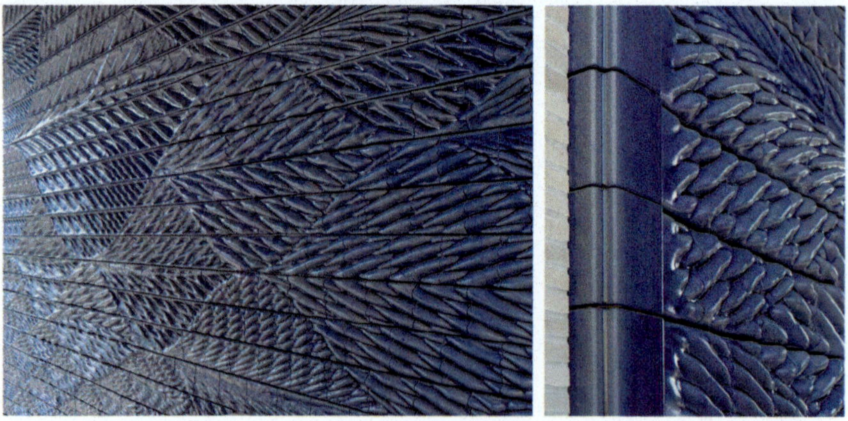

Fig. 6 The New Delft Blue project close-up pictures (R. De Vecchi photographer)

4 Ceramic House

According to Studio RAP, The Ceramic House reshapes architectural expression by seamlessly blending tradition and innovation, reintroducing bespoke details to the realm of architecture. The studio's innovative approach to contemporary architectural design uses state-of-the-art digital fabrication processes conceived in its own manufacturing premises to print each bespoke 3D-printed ceramic tile. This enriches the famous shopping street's existing historic architecture with a new design language. The resulting design, achieved through algorithmic design, 3D-printed ceramics, and artisanal glazing, is exciting and textural, inspired by the craft of knitting garments (see Figs. 7 and 8).

4.1 Innovating Algorithmic Design

Using digital design algorithms designed in-house, the Ceramic House explores a reinterpretation of the decorative qualities and design vocabulary of glazed ceramics in the historical city of Amsterdam. Studio RAP has replicated the silhouette of the original facade, continuing the characteristic tripartite structure of the street and maintaining the overall character of the site. The scale, size, type, and color of the

Fig. 7 Close up picture of the Ceramic House project (R. De Vecchi photographer)

Fig. 8 Picture of the plinth of the Ceramic House project (R. De Vecchi photographer)

ornaments and materials are all carefully synced with the neighbouring buildings to allow a seamless integration of traditional and contemporary architecture. The design of the façade features intricate layers inspired by textiles—elegant creases, interlooping yarns, and stitch patterns. Its organic, wave-like quality changes as viewers approach the design from different angles. Gradually, as the line of vision moves, new elements within the custom ceramic tiles are unveiled, resulting in a luxury boutique that harmonises with its historical environment while standing out among the surrounding buildings.

4.2 *Pioneering 3D-Printed Ceramic Tiles*

Drawing inspiration from the Rijksmuseum, the national museum of the Netherlands celebrated for its diverse ceramic collection from across the globe, Studio RAP employs their in-house large-scale 3D-printing technology to realise highly differentiated and algorithmically designed details. This innovative approach underscores Studio RAP's commitment to pushing the boundaries of architectural design Studio RAP has developed a distinctive digital fabrication process over several years, including the creation of advanced robotic systems. The team of architects collaborates seamlessly with these robots to precisely craft intricate ceramic designs, showcasing the studio's commitment to excellence in architectural design. (Fig. 9) At street level, the facade features large 3D-printed ceramic tiles, approximately 40 x 20 cm each, glazed in pearl white with a subtle touch of yellow by Royal Tichelaar. These tiles are designed to be visually expressive at eye level, seamlessly transitioning to a flush alignment as they meet the ground, creating a harmonious and detailed aesthetic. On the higher floors, the building facade showcases 3D-printed bricks glazed in three distinct shades of red. The bricks are composed alongside the original masonry cross bond and have abstract ornamentation that fades as they travel higher. These bricks are put into laser-cut stainless steel cassettes, referring to the original flushes (Fig. 10).

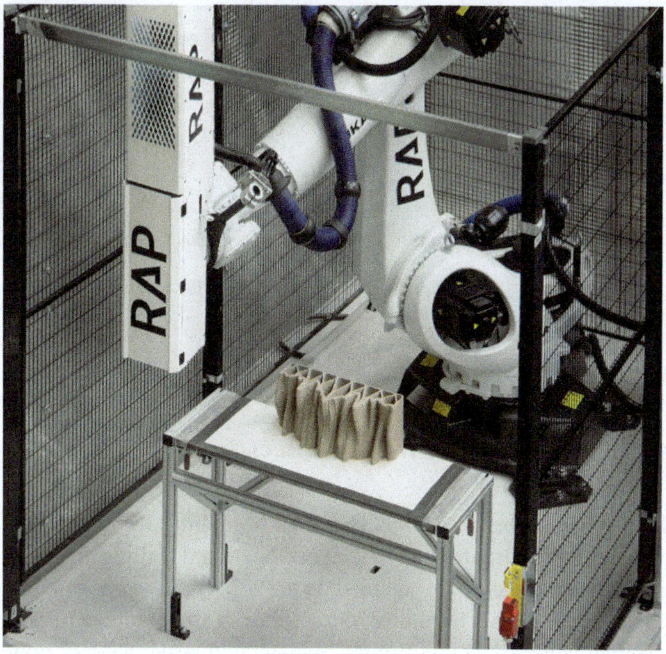

Fig. 9 Picture of one of the 3d printing clay robots at Studio RAP

Fig. 10 Picture of the bricks of the Ceramic House project (R. De Vecchi photographer)

5 Future is Now

5.1 New Added Values of the Architect

For the three projects the client was specifically looking for a different approach to realize something extraordinary and were open and willing to adapt to a new digital workflow to realize it. This opened up the opportunity to position the architect in a central role and as such adding extra value in the creation of a streamlined digital workflow while touching a more fundamental issue, namely how architects can strengthen their position from which they can claim and protect their creativity and aesthetic concerns under capitalist conditions.

Studio RAP

Studio RAP is an award-winning architectural design and fabrication firm based in Rotterdam (NL). The studio combines the power of computational design with innovative digital fabrication methods. Their team consist of architects, designers, roboticist, programmers and researchers. Their mission is to inspire people everywhere in the world by bringing innovation to architecture. In this way the studio aspires to enrich the world with a new architecture language through the power of emerging digital technologies, breaking barriers by innovating the way the world is envisioned, built and experienced. Making the world more exciting and diverse.

Reference

1. Samuel F (2018) Why Architects matter: evidencing and communicating the value of architects. Routledge

Lucas ter Hall is an architect and co-founder of Studio RAP. He studied architecture at the Technical University in Delft and graduated Cum Laude in 2014. After graduation he started Studio RAP together with Wessel van Beerendonk. He worked on projects like New Delft Blue, Ceramic House, The Float, Theatre Zuidplein and Skilledin Office.

Chapter 15
Superframe: A Novel Construction System that Combines Planar Panels with 3D-Printed Frames Made from Recycled Thermoplastic Materials

Pierpaolo Ruttico, Federico Bordoni, and Simone Maccagnan

Abstract The authors of this chapter introduce Superframe, a system developed to create cost-effective free-form claddings. The innovation of the system is centered around the capability to combine planar n-gonal panels with twisted-edge frames, effectively addressing the discretization of intricate surfaces. This solution is made possible through a 3D printing process that shapes the frames according to the required geometries, utilizing recycled plastic through Robotic Additive Moulding (RAM) technology. The 3D-printed frames can be paired with various interior or exterior cladding materials, offering numerous configuration possibilities and granting designers unparalleled creative freedom. In a broader sense, Superframe represents an approach to complex geometries that is mindful of the challenges—or field of forces—of our time, tied to conscious resource utilization, and leveraging the possibilities offered by computational design.

Keywords New construction system · Additive manufacturing · Mass-customization · Double-curved surface discretization

---

P. Ruttico (✉) · F. Bordoni
INDEXLAB, Politecnico di Milano Via Gaetano Previati 1c, Lecco, Italy
e-mail: pierpaolo.ruttico@polimi.it

F. Bordoni
e-mail: federico.bordoni@polimi.it

S. Maccagnan
eXgineering, Via Cantonale, Manno, Switzerland

Gimac, Via roma 5, Castronno, VA, Italy

S. Maccagnan
e-mail: simone.maccagnan@gimac.com; sm1@exgineering.com

P. Ruttico (ed.), *Coding Architecture*, Digital Innovations in Architecture, Engineering and Construction, https://doi.org/10.1007/978-3-031-47913-7_15

1 Introduction. The Quest of a Good Discretization

In his "Notes on the Synthesis of Form", Christopher Alexander claims that form is the direct consequence of a compromise, a negotiation, with its external context: the ultimate result of a struggle in a field of forces [1]. Form is immersed in an irregular world that tries to compensate for its own irregularities by fitting itself to them. It is tempting, both for the inexperienced designer and the casual wanderer alike, to dismiss the topic of form-finding as a mere technical matter: a kind of data analytics challenge; to derive good form is a process aimed at responding coherently to the field of forces that surrounds it, so the better the description of this field, the more precise a "counter-form" can be derived.

This form-fitting problem is the standard action-reaction equation that, once these irregularities are found, employs form to "counteract" them.

Regrettably, for designers at large, faithfully transcribing and mirroring this field in its current state yields neither beauty nor inherent credibility. That is to say that, somewhat expectedly, the most literate copy of reality still misses the glimmer of the truthful. There seems to be a further translation that is necessary to give credibility to form. This translation intuitively resonates with the discretization strategy employed to craft a certain form in our reality and the results achieved in its production. The strategy will also visually codify the overall plausibility of the effort undertaken to manufacture that form and its general goodness of fit in its context.

Good form is the result of a translation of a context made of infinite forces into a simpler model which, taking into account only a part of these forces and ignoring some others, embodies a principle on how to understand and ultimately represent this context. This principle comes with an embedded discretization bias that will be less important the less evident it is to the eye. Achieving a good discretization of form relies on the contamination with the present, its technological realities, its construction industries, its socio-cultural milieu. Good discretization, in architecture as an example, is a fundamental aspect that not only systematizes but also classifies form quality, since the leap of pragmatism separating geometric form-finding and physical form-fitting is key to its goodness of fit. As noted by Pottmann, Brell-Cockan and Wallner in Discrete Surfaces for Architectural Design: "The reasons for using nearly developable surfaces are rooted in manufacturing and fabrication. In view of the large scale on which surfaces in architecture have to be built, it is obvious that the choice of the fabrication technique has an influence on the surface representation and on the design principle" [4]. That is to say, discretization quality has the power to influence all the design process from the beginning, from its representation to its fabrication. That is why scripting—including both the codification of a design principle for form-finding and the part of form-fabrication in a single and coordinated algorithm—provides a more effective overview of the whole process, from concept to production. The feedback loops between design and fabrication visually shed light on the perks and drawbacks of the chosen discretization strategy.

The authors intend to present the results of INDEXLAB and Gimac-eXgineering's ongoing research on pellet-extruder based Additive Manufacturing through the documentation of three prototypes concerning the construction of discretized complex curvature surfaces. These examples are meant to both document the processes used and the role of INDEXLAB in the innovation of the digital workflows under the light of the fabrication challenges for the construction industry.

2 Form Generation Methods and Fabrication Challenges

The Superframe construction system was developed to address a specific need: streamlining production timelines and material usage, resulting in cost savings, particularly for crafting free-form cladding structures. In architecture, construction elements are usually manufactured in two-dimensional, flat forms, but architects and designers often aim to create more dynamic, curved buildings. To achieve this, one approach is to break down complex surfaces into simpler planar elements, with the complexity shifted to the underlying structure. This substructure can be 3D-printed with twisted sides to accommodate the curved surfaces effectively. This is where the concept of Superframe comes into play—a framework that enables the efficient creation of intricate spatial designs using minimal material, thus offering a cost-effective solution (Fig. 1).

The development of Superframe by Indexlab-Gimac-eXgineering has progressed through three stages. The initial version (v1) was designed as a lightweight facade system for special buildings requiring curved claddings, such as shelters or uniquely designed building elements. Several mock-ups were created, featuring different cladding materials mechanically attached to the 3D-printed frame.

The second version (v2) introduced a novel concept: directly attaching panels to frames by depositing thermoplastic material within the panel's cavities. This method

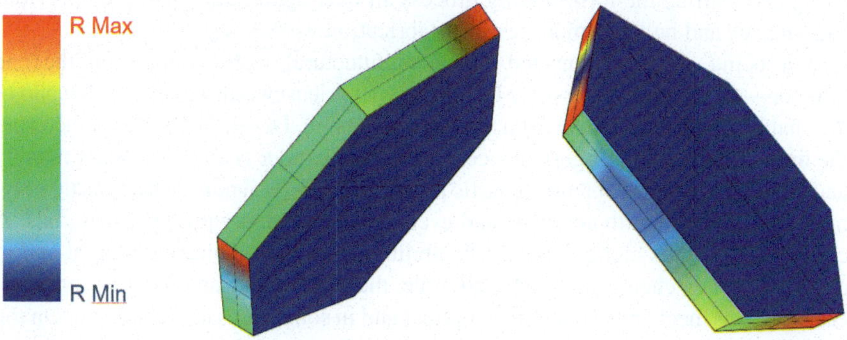

Fig. 1 A diagram showing the Superframe concept: a system that combines flat panels with a substructure designed to address a technical challenge: the torsion that arises between adjacent profiles in the frames to which the panels are attached

has been patented by Gimac and Indexlab, and demonstrated significant innovation potential. Superframe v2 was successfully applied to clad the "Urban Lounge" installation, a Pininfarina Architecture project showcased at RoGuiltlessPlastic 2021 in Milan, highlighting the practical application of the system. All of this within a very short timeframe of just two months, during which the transition from a program to a previously non-existent physical object occurred. This involved the conception of the form, study of the topology, development of technology, technique, and process, as well as obtaining approvals for implementation in a public space along with related logistical matters.

The third version of Superframe (v3) includes prefabrication enhancements, such as CNC-drilling techniques for inter-frame fastening, and software improvements for managing assembly tolerances and ensuring consistent gaps between cladding panels, now in a "cassette" configuration. Demonstrative prototypes of this enhanced version have been developed, marking a notable step forward in the Superframe's evolution.

Since all of them are indeed successive fine-tunings of strategies aimed at solving the same problem—the fabrication of Discretized Complex Curvature Surface—we refer to them for concision respectively as iterations (v1, v2, v3) of the same project, Superframe. INDEXLAB's will to create a biunivocal translation between fabrication process and form-generation via the medium of computation is the main fil-rouge between all these iterations. This resonates with the fact that the use of parametrization and scripting allows to rapidly commute—to transfer—the knowledge matured prototyping a certain solution directly towards a consolidated industrial process, igniting innovation and a possible re-framing of the role of automation for the construction industry.

Specific Design for Additive Manufacturing (DfAM) techniques and tools are required to fulfill the request from the main stakeholders. The goal is to employ robotic manufacturing processes (RAM) to automate the construction workflow in order to circumscribe manual labor to the sole on-site assembly of prefabricated modules. DfAM processes are key to fast-prototyping, receiving feedback, simulating and starting the form-finding process all over again, capitalizing on previous experiences and building more resilient fabrication workflows.

Not being strictly connected to the architectural world, details on the end effector—AKA the extruder—are left out, but it's at least worth to mention that part of the enabling substrate comes from the extruder design. Developed for manufacturing the first microinvasive surgery devices, the microextruder is able to process reliably and precisely any kind of thermoplastic polymer and several non-thermoplastic polymers too, in blend with additives and even in co-extrusion with other materials. A core feature of this end-effector is the ability to preserve molecular weight and to condition, when chemo-physically possible, the crystallinity and the deriving properties such as mechanical, electrical, optical and in some cases also chemical. On top of that, other topological aspects can be applied, like foaming, coextrusion and cross-sectional features on the deposition cord that doesn't necessarily need to be a solid circular section, but could be anything from circular to star-shaped, solid, hollow, foamed or multimaterial, allowing not just cosmetic effect, but also highly functionalized structures, i.e. conductive paths within the construction, allowing to bring signals or power into or out from the otherwise inaccessible core of the structure. Sensorised or active structures becomes then real.

Thanks to such a high level of sophistication, this end-effector allows to get the best out of damaged polymers, such as post-consumer or ocean plastics after mechanical recycling, in other words, materials that does not necessarily have much more to offer before a chemical reprocessing.

2.1 Superframe v1

Superframe v1 originates from the need to address the lack of a simple and effective construction method for prefabricating and assembling free-form claddings. One of the most promising applications lies in the growing market of glamping structures, compact designs preferably incorporating organic shapes. Another highlighted application is for free-form railings, elements not subject to overly restrictive regulations, and finally, semi-transparent partitions, also serving as sunscreens. The initial investigation focuses on discretization methods. Positive Gaussian curvature is initially chosen, along with three subsequent patterns: a hexagonal mesh, a trapezoidal pattern resembling a "planar quad," and lastly, one based on a triangular mesh. All initial frames produced feature a 3D print with one or two lateral perimeters and an interior zig-zag lattice. ABS-extruded modules were subsequently chemically or mechanically bonded to various materials: externally with carbon fiber and Kevlar laminates, and internally with wood for the "glamping" building model; high-pressure laminates and metals for free-form railings; innovative textile materials with a thermoplastic matrix and light-permeable properties for sunscreen elements (Figs. 2, 3, 4, 5 and 6).

Fig. 2 A rendering displaying a concept for a "Glamping" building enclosure. The enclosure takes on a free-form shape, discretized into planar n-gonal elements, from which a small module is extracted as a demonstrator

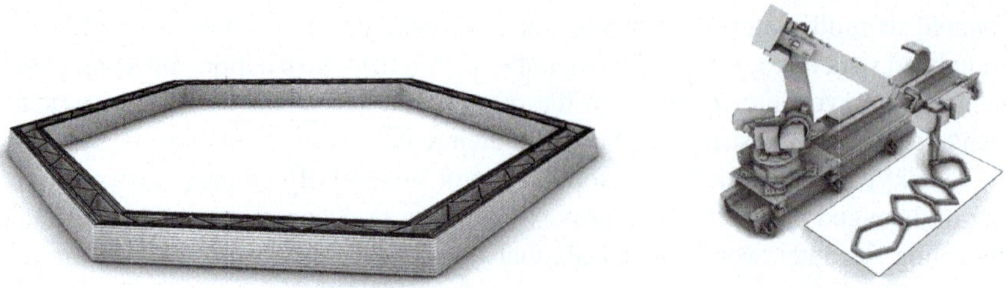

Fig. 3 Rendering on the left showcase a cross-section of a frame with two perimeter layers and a zig-zag infill; on the right, a robot sequentially printing panels

Fig. 4 Photos of the initial Superframe v1 mockup, exhibited at the 2020 SAIE fair. The prototype features a 3D printed ABS frame mixed with glass fibers, externally laminated with three different types of carbon and kevlar-based fabric materials, and internally covered in wood. A layer of rock wool is placed within the cavity. The central module includes a double glass pane, simulating a glazed opening

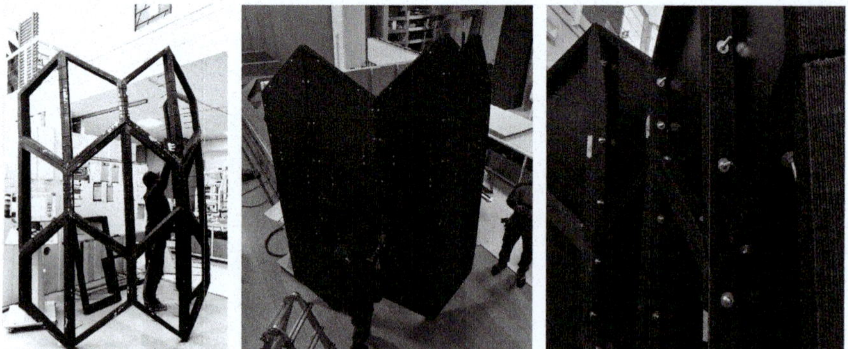

Fig. 5 Photos of the second Superframe v1 mockup, intended for a building with flowing curves' railing. This mockup features ABS frames mixed with fiberglass and clad in high-pressure laminated material, riveted to the frames

Fig. 6 Photos of the third and fourth Superframe v1 mockups, prototypes combining superframe with two other types of cladding: one semi-transparent thermoplastic material suitable for partitions or sunscreens (on the left), and the other (on the right) coated in gold-effect metal, showcasing the system's versatility

2.2 Superframe v2

The evolution in Superframe version 2 stems from a request by Pininfarina Architecture to co-design an installation at the "Museo Nazionale della scienza e della tecnologia—Leonardo da Vinci, Milano", a project later named "The Urban Lounge". This involves a small-scale free-form enclosure intended to cover a sophisticated air treatment unit with a 2 × 2 × 5 m bounding box. The rationale behind the project's development is intrinsically linked to a parametric approach that takes into account all influencing factors: foremost among these are time, costs, sustainability, manufacturability, and the effectiveness and efficiency of the system. The goal of every project beginning with unknown design boundaries is always the same, and always very ambitious: to perfect a simple yet effective, all-encompassing algorithmic workflow to solve a problem, with an eye on the manufacturing process from the start, as if it was already embedded in the code. This kind of geometric discretization is usually a consequence of long iterative trial-and-error processes which cannot ensure the mesh is optimized for fabrication if not through formalizing some constraints; in this case, the triangulation of the mesh needs to obtain the maximum number of equilateral triangles. A constant back-and-forth to tweak design aspirations to structural, production and timing constraints results in the development of a re-dimensioned Grasshopper3D script, aimed at triangulating a surface with certain design boundaries and ensuring that both the maximum draft and the vertex angles of each mesh triangle be adherent to the manufacturing limits of the 3D-printing technology (Fig. 7).

INDEXLAB prepared the manufacturing files, each associated with printing time and material cost. The Grasshopper3D script converts the 3D shapes of each frame into a continuous toolpath for the 3D-printing process, allowing control over a set of printing features through their parametrization. The two most innovative aspects of the Superframe v2 system, though, are the geometry of the interface between

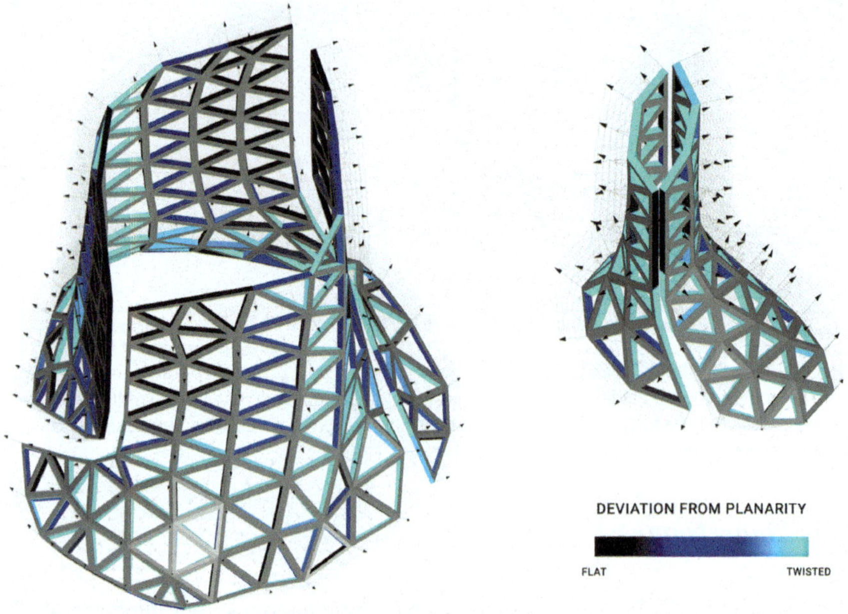

Fig. 7 Diagram representing a set of triangular modules, each distinct, with varying degrees of torsion between adjacent faces. Color variation signifies planar deviation—black and gray indicating flat faces, cyan depicting maximal twisting. Vectors depict normals to the doubly curved surface at the vertices of the triangulated triangles' discretization

adjacent ABS panels that ensures structural stability and an even distribution of the stresses along their edges, as well as the interface between the ABS extrusion portion and the aluminum module covering it.

The first aspect is solved designing ruled interfaces for the edges of each panel, which could be effectively discretized granting good adherence between panels also in the realm of the high tolerances of classic Additive Manufacturing through pellet extrusion. The intuition lies in the fact that, if each layer of the panel is rotated and its shape distorted relative to the panel's base, it is always possible to achieve a clean surface of adhesion between adjacent panels. Even though the broad tolerances dictate that no glue can be applied to connect them, they can be easily dry fastened on-site using M8 screws and washers. This is of paramount importance, since every assembly phase on-site has to be managed in consideration of the museum's strict regulations and ensuring that the least amount of heavy machinery is used inside the museum itself. Furthermore, the ruled interface solution (the twisted surface) retroactively rules over the triangulation of the mesh of the beginning, piping surface discretization quality, mechanical strains in each edge and assembly of the panels, all in one coherent effort.

In essence, the ruled interface solution acts as a comprehensive framework that interconnects various aspects of the project—from mesh triangulation and surface

Fig. 8 Photos of the 3D printing process

discretization to mechanical strain management and panel assembly. It demonstrates how a holistic and well-thought-out approach can result in a harmonious and functional outcome.

Prototyping a solution to the second aspect, which seems—at first—of a smaller order of magnitude compared to the first one, is instead crucial for the success of the whole operation, since gripping each aluminum module manually would extend construction timings and be very problematic to center on top of its underlying ABS extrusion. The insight, in this case, is to use honeycomb sandwiched aluminum modules and to mill one side with a dovetail-shaped tool. The sandwich panel limits the overall cost and weight of the aluminum cladding, since it is fabricated using thinner aluminum sheets, while ensuring comparable stiffness to the panel. The superficial milling of the panel, instead, creating a socket which is filled with the first layer of plastic extrusion, secures the ABS to the aluminum—also neutralizing the effects of solidifying plastic shrinking of the fused plastic caused by its temperature drop from extrusion to ambient temperature (Fig. 8).

This method has been patented and represent an innovative solution in the field of additive manufacturing.

From a production optimization standpoint, all the necessary information for the assembly and the 3D printing process are integrated into every single aluminum module. Each sheet acts as a base and a blueprint for the components of the Superframe v2 (Fig. 9).

Each nested panel possesses a distinct ID and neighboring tags, facilitating straightforward spatial identification within the installation area. INDEXLAB coordinates both the off-site production, undertaken by Gimac-eXgineering, and the assembly of the installation, performed by Nieder. The on-site works are coordinated also with the other installation teams, ensuring that all operations proceed smoothly, and the manufactured panels are pre-assembled off-site into clusters to reduce on-site costs and delays (Figs. 10 and 11).

Fig. 9 Photo of the macro modules assembly phase of Superframe version 2. During this stage, the modules are manually aligned, secured with clamps, drilled and screw-fastened in pairs

Fig. 10 Photo of the on-site assembly process

2.3 Superframe v3

From the experience of Superframe v2, some insights and newer ways to think about the whole manufacturing process arise. Two facts are deemed fundamental to preserve: firstly, the fact that Superframe v2 entails a customary three-axis print, allowing simpler and faster fabrication—possibly also with relatively small three-axis machines; secondly, the calculations to generate mesh polygonalisation possibilities are fast and relatively painless using an ordinary desktop machine.

The employing of recycled post-consumer materials is also a key strength, as repurposing ABS pellets into rigid frames used as structure for a façade not only

Fig. 11 Photo with no filters applied of the final installation at the Museum of Science and Technology in Milan, as part of RoGuiltlessPlastic 2021

dignifies and saves a material form the landfill giving it a third life, but also contributes to expand research about synergies between seemingly distant economic sectors or other cross-industry applications. The material of Superframe v3 ought to be exchangeable for a yet non-characterized material, this in order to expose the research activity to a more responsible use of material, for example using self-generated scrap plastic or short distance generated plastic from ordinary human activities, this of course doesn't have any sort of datasheet, but still it could be characterized on the go, by mean of simple analytical equipment like: FTIR, DSC, TGA or simply by STA and FTIR, this could serve as a tester for large-scale additive moulding (AKA 3D printing, additive manufacturing, etc.).

As the theme of metal replacement becomes more and more central in the aerospace and automotive industries, the construction industry follows along, benefiting from contemporary prototyping experiments and the study at the scale of architecture of how some recycled plastic blends mechanically behave, resist to fire and water, endure great heat excursion. From Gimac-eXgineering's pellet-tech and industrial extrusion knowledge, a special blend is selected—a blend of, not necessarily predefined, post-consumer waste plastics is composed in order to contain shrinkage and warping thanks to the introduction of fractions of waste containing elastomeric functional groups. Since the fastest production of the panels is desirable, Superframe v3's modules are printed at 100 mm/s, thus entailing faster collapse rates of the material and the use of cooling media (Fig. 12).

Thanks to this deposition system and new materials blend, there is no need to use any adhesive agent on the print bed, the shrinking and warping effects can easily be contained without the use of a thermostatic cell. The use of materials that feature less warp led to the possibility of automating another activity—previously done manually

Fig. 12 On the left, a photo shows the first layer extrusion process in Superframe v3. The base has references for placing cassette-style metal modules for the façade system. On the right, a picture depicts a robot drilling the frame with precise angles for accurate hole alignment, ensuring efficient assembly

and on-site: the milling of the holes necessary to fasten two panels together, as well as the milling of the faces for improving the tolerances magnitude. The same approach to meshing and evaluating overhangs and draft angles of each Superframe v2 panel is thus re-engineered in Superframe v3 to comprise the milling of the holes using a second robotic arm. The holes need to be approached by a milling spindle—without the spindle colliding onto the underlying print bed—and ensuring that, as the spindle approaches and impacts the panel, this adheres enough to the print bed to prevent vibrations that would de-center the milled hole. Featuring a very ordinary milling spindle, holes are drilled to accommodate the M8 screws and washers necessary to fasten all the pieces together (Figs. 13, 14 and 15).

Fig. 13 On the left, a photo demonstrates improved panel anchoring in Superframe v3 compared to v2. Chamfered angles prevent over-extrusion, enhancing tolerances. Also, notice metal tabs on cassettes aligning with frame tabs. On the right, see consistent gap quality between metal cladding cassettes despite panel angle variations

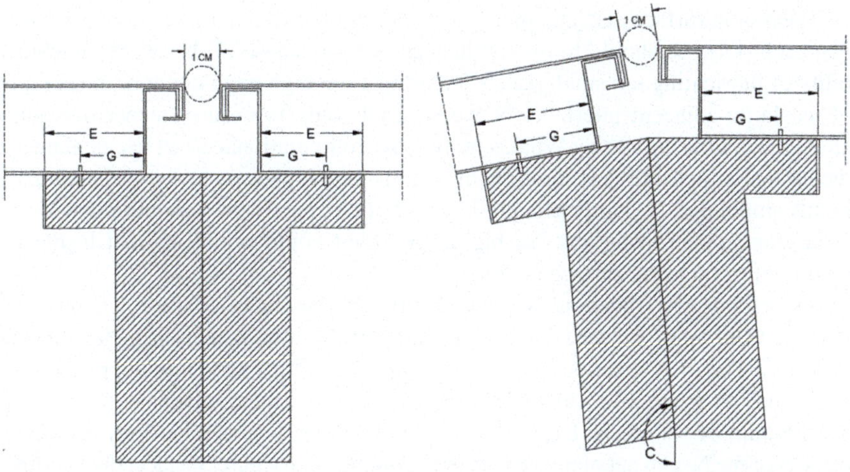

Fig. 14 Diagram showing the variation in panel positioning based on the inclination of the supporting frames. Panel seams are constant while distances E and G are variables dependent on the angle of the frames C

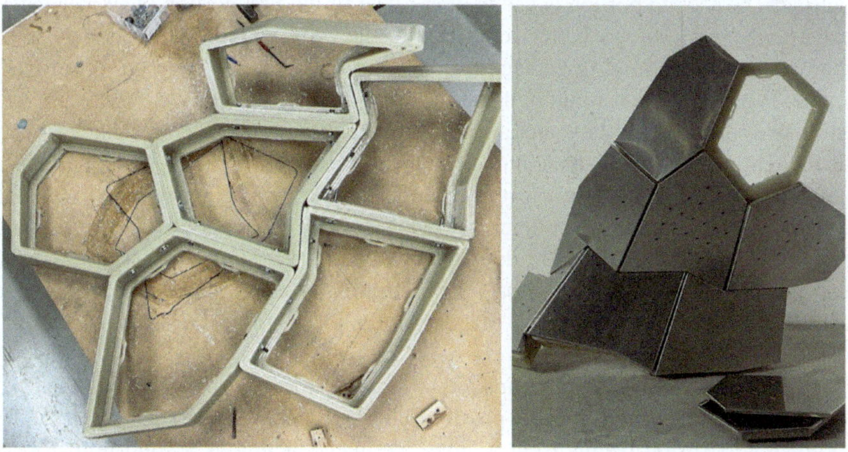

Fig. 15 Photo of Superframe v3 modules assembled together and subsequently covered with a cassette-style metal panel system

The main improvement of Superframe v3 nonetheless is to free the meshing computation from the initial constraint of a solely triangular polygonalisation and to project it towards an N-gon surface discretization. This cannot be achieved without stating certain boundaries of "concern" or application for the definition of the algorithm.

The N-gon surface discretization has to come with a set maximum number of faces that each polygon should have—in this case six—because of the overall implausibility of fabricating too small panels featuring a too high edge-count or the creation of too large adjacent interfaces between panels only to ensure steeper overhangs along the double curvature. What is more—the computational effort to calculate and the aesthetic pleasure that comes with seeing—a complex surface tessellated using highly populated N-gon panels clashes with the economical reasoning that would want every panel to be just "as big as the robot could print" without losing the feeling of the starting surface geometry.

Superframe v3 is thus conceived as an (hexa) N-gon frame structure, a Discretized Complex Curvature Surface featuring planar portions always measuring less than 60 × 60 cm. The discretization uses evolutionary algorithms and Grasshopper3D add-ons—such as Kangaroo—to find the best-fit scenario using planar polygons with a maximum of six edges. Every Superframe v3 panel is generated from an atomic portion of the N-gon net of the Discretized Complex Curvature Surface: the "outline" of the panel.

Looking at each panel from its side, it is easy to notice that its geometry is generated from the superposition of two portions. The first portion measures around two thirds of the piece in height and features a three-perimeters inner offset of the outline. The second portion is yet another offset of the first portion (a six-perimeter extrusion) which allows better adherence to the printing table without the use of glue and creates a kind of setback to place the cladding correctly on top of each frame. Without the constraint of using the honeycomb aluminum module to prevent excessive shrinkage of the material and attaching the two together, the cladding can be thought of as a simple dry-mounted element on top of the panel. It is important to note that, since every frame—just as in Superframe v2—is geometrically a twisted extrusion of its pertaining outline, the setback's face is parallel to the outline and planar, yet always twisted and deformed compared to the outline. It is precisely this planar distortion of the setback that ensures an equal spacing between each aluminum module is retained and that each module be centered according to the 3D model (Fig. 16).

Every frame's toolpath is optimized so as to feature the least possible material waste and to allow the cleanest possible interface between adjacent frames. The toolpath's extrusion begins from a "mid" perimeter acting as a solid infill for an external perimeter, fitted to the outer edges of the frame, and for an internal perimeter, which is fitted to the setback's depth at first and then continues contiguously to the mid perimeter for the remaining two thirds of the frame (Figs. 17, 18 and 19).

3 Results and Further Developments

The "minimum working example" of Superframe v3 has been achieved with developments and ameliorations in the robotic cell setup (the use in combination of additive and subtractive processes in a sole fabrication workflow), the knowledge concerning

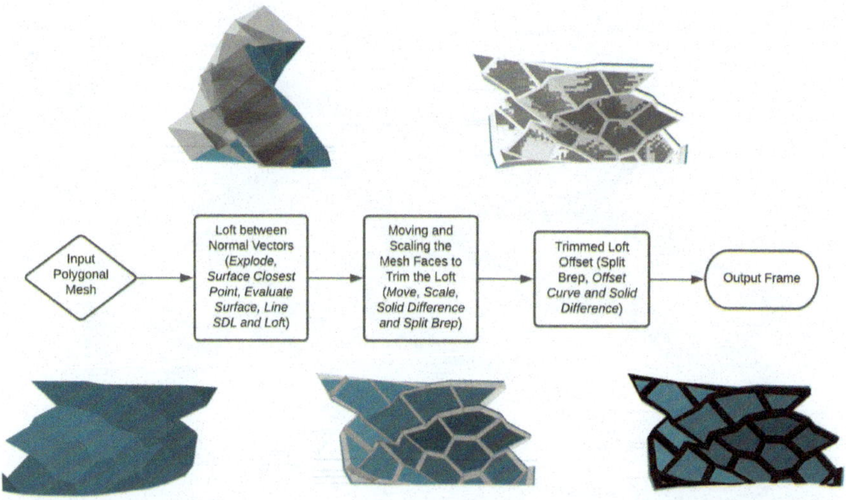

Fig. 16 Diagram showing the generation of the frame from the edges of the polygonal mesh. To find the inner and outer surfaces of the supporting structure, each face of the polygonal mesh must be moved internally following its normal vector and scaled to intersect and to cut the created loft twice. The surfaces are moved firstly by a distance equal to the panel thickness and then by a distance equal to the frame depth (the frame depth and the distance between the outer surface and the frame are two variables of the parametric script). This is carried out using two Grasshopper transformation functions, "Move" and "Scale", and two intersection functions, "Solid Difference" and "Split Brep". This results in a trimmed loft representing the edge surfaces of the frames of the supporting structure

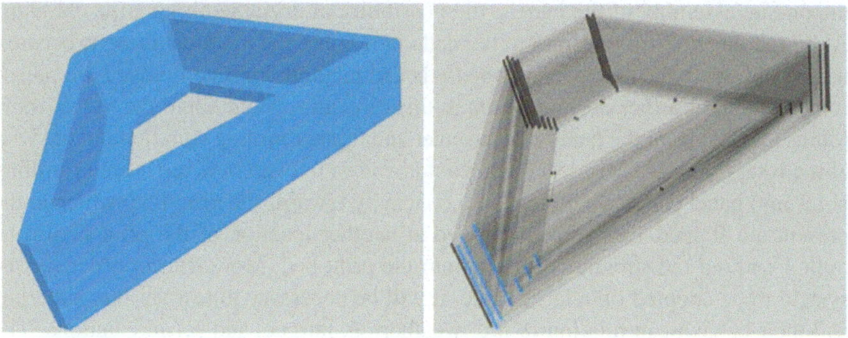

Fig. 17 Diagram representing the frame contouring. Its geometry combines two parts: one covers around two-thirds of the height, with a three-perimeter inner offset. The other, a six-perimeter extrusion, aids adhesion to the print bed and facilitates proper cladding positioning

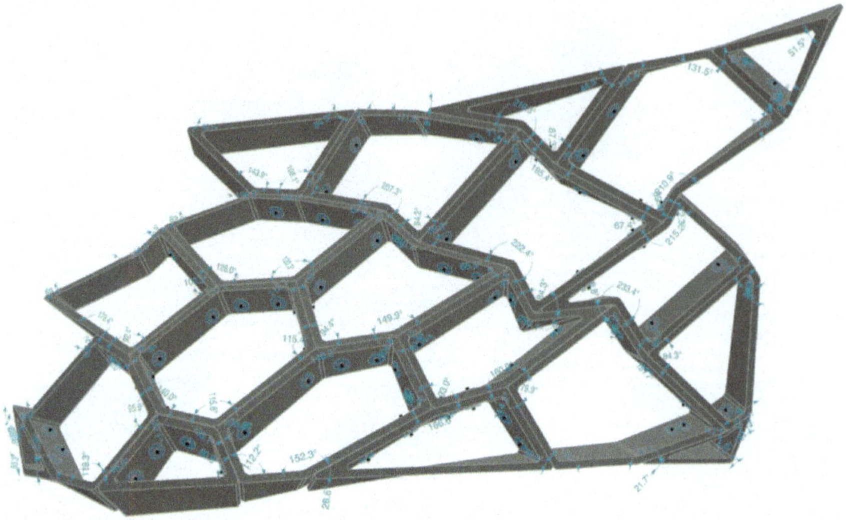

Fig. 18 Diagram showing the bolt positioning based on node angles. The manipulation of data lists and trees pairs a vertex of the polygonal mesh (associated with a maximum of three angles) with each midpoint of the edge lines. Considering the minimum angle at a given node, the midpoints move from the edge lines to specific locations on the contact surfaces. The script defines five empirically derived angle ranges ($a \geq 90°$, $60° \leq a < 90°$, $45° < a < 60°$, $36° \leq a \leq 45°$, $a < 36°$), corresponding to five distinct edge-to-hole distances

previous experiments with triangular, and quadrilateral meshes, the will to speed up the on-site assembly of the panels, the possibility of dry-assemblying the cladding. Although there is still progress to be made in refining the manufacturing processes, it can confidently be asserted that the tools and methodologies employed are poised for scalability. Further developments in this direction would entail the use of more robots to batch print the panels, of stronger and thinner milling spindles and of swivelling rotary tables. These last ones would be used to overcome the issue of milling holes into panel edges which feature a concavity towards the print bed, which at the present are flipped 180° and positioned at another location of the print bed to be milled without the spindle colliding onto the print bed. Moreover, in order for the system to be adopted on a large scale, it will be necessary to test various materials and their behavior in relation to fire, weathering, fatigue, and seismic actions. The journey towards enhanced industrial applicability continues.

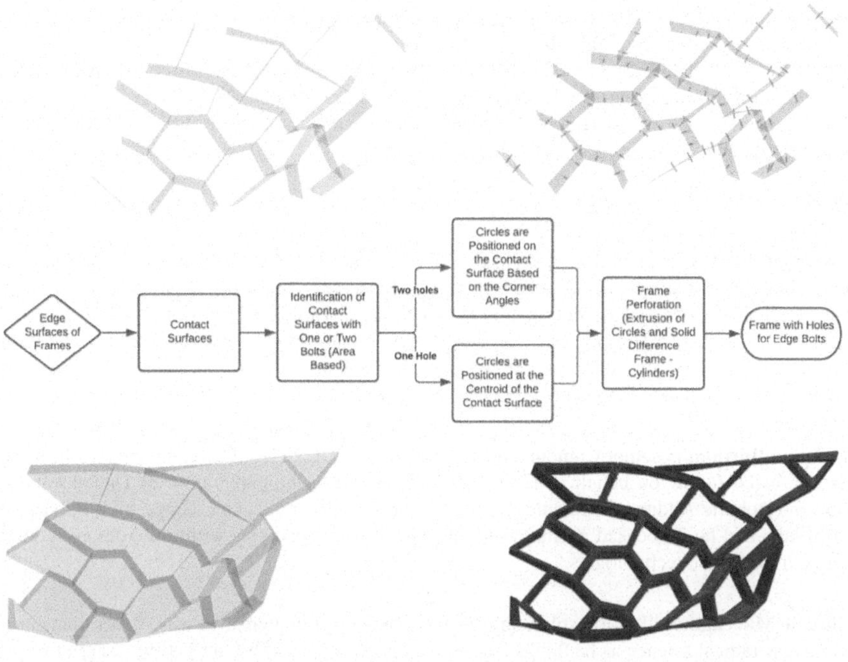

Fig. 19 Diagram showing the edge bolts positioning. All contact surfaces are perforated through these steps: (1) constructing circles with a radius equal to the bolt radius at each point on the contact areas with Circle CNR, (2) extruding them in both directions by a distance equal to half the bolt length, and (3) applying Solid Difference between the frames and the created cylinders. The main inputs of this parametric algorithm include the hole radius, half the bolt length, and the edge-hole distances per range of angles

References

1. Alexander C (1964) Notes on the synthesis of form. Harvard University Press, Cambridge, Massachusetts
2. Kepes G (1965) Structure in art and science, 4th edn. George Brazieller
3. Pottmann H, Brell-Cokcan S, Wallner J (2006) Discrete surfaces for architectural design. Curves and surfaces: Avignon 2006, pp 213–234
4. Pottmann H, Brell-Cokcan S, Wallner J (2006) Discrete surfaces for architectural design. In: Chenin P, Lyche T, Schumaker LL (eds) Curves and surface design: Avignon 2006. Nashboro Press, pp 213–234
5. Liu Y, Wang W (2009) A note on planar hexagonal meshes. In: Nonlinear computational geometry. Springer, pp 221–233
6. Liu Y, Xu W, Wang J, Zhu L, Guo B, Chen F, Wang G (2011) General planar quadrilateral mesh design using conjugate direction field. ACM Trans Graph (Proc SIGGRAPH Asia) 30(140):1–10
7. Pellis D, Kilian M, Wang H, Jiang C, Müller C, Pottmann H (2021) Architectural freeform surfaces designed for cost-effective panelling through mold re-use. In: Proceedings of the Advances Architectural Geometry, Paris, France, May 2021
8. Pluta K, Edelstein M, Vaxman A, Ben-Chen M (2021) PH-CPF: planar hexagonal meshing using coordinate power fields. ACM Trans Graph 40(4). (Article 156)

9. Poranne R, Ovreiu E, Gotsman C (2013) Interactive planarization and optimization of 3D meshes. Comput Graph Forum 32(1):152–163
10. Pottman H, Schiftner A, Bo P, Schmiedhofer H, Wang W, Baldassini N, Wallner J (2008) Freeform surfaces from single curved panels. ACM Trans Graph 27(3):76
11. Pottmann H (2013) Architectural geometry and fabrication-aware design. Nexus Netw J 15:195–208

Pierpaolo Ruttico is an architect-engineer and researcher at Politecnico di Milano, specializing in complex constructions and advanced production processes. He previously worked in New York City for Pelli Clarke & Partners, then earned a Ph.D. in Building Engineering, with a specialization in non-standard geometries. He is the founder and director of INDEXLAB, a multidisciplinary research group established in 2010. INDEXLAB focuses on contemporary architectural languages and pioneers innovative processes in digital fabrication, AI and robotics.

Federico Bordoni is Senior Computational Designer at INDEXLAB, Lecco. In this role, he serves as a research fellow on behalf of Gimac S.r.l., focusing on the exploration of Digital Fabrication and Robotic Additive Moulding. He holds a Master of Architecture degree from Politecnico di Milano and he is focused on innovating manufacturing paradigms within the design industry through disruptive technologies.

Simone Maccagnan likes to explore the existing knowledge in a creative way, rather than generating new radical science as he thinks that in this historical period we are getting faster in forgetting existing knowledge, that costed pollution, loss of biodiversity and CO_2 generation. Thechnosopher, eco-anxious and in love with Humans, he is business development manager at Gimac, global leader in the development of polymer processing equipment for the production of medical devices. A bit less than 10 years ago, he restarted to put momentum on the research initiated by his Father Giorgio about 40 years ago, when, in the early 80's, he made a pellets based additive manufacturing system using his 10 mm micro extruder and a cartesian milling machine for solving manufacturing problems in the production of extrusion tooling.

SPRINGER NATURE

GPSR Compliance

The European Union's (EU) General Product Safety Regulation (GPSR) is a set of rules that requires consumer products to be safe and our obligations to ensure this.

If you have any concerns about our products, you can contact us on ProductSafety@springernature.com

In case Publisher is established outside the EU, the EU authorized representative is:

Springer Nature Customer Service Center GmbH
Europaplatz 3
69115 Heidelberg, Germany

The manufacturer's authorised representative in the EU is Springer Nature Customer Service Centre GmbH, Europaplatz 3, 69115 Heidelberg, Germany. If you have any concerns regarding our products, please contact ProductSafety@springernature.com

Printed and bound by CPI Group (UK) Ltd, Croydon, CR0 4YY
01/10/2026
02225866-0008